ANOTHER PLACE ANOTHER TIME

ELAINE BAKER

Golden Age Publishers

First published by Golden Age Publishers in 2020
ISBN: 978-1-8382756-2-4
Copyright © Elaine Baker, 2020

ACKNOWLEDGMENTS

I would like to thank all the people mentioned for their contribution. For those from whom permission has not been formally requested or granted, I sincerely hope they would have approved. This book captures memories of a past era, but sadly many of those friends and colleagues are no longer with us. Some of the photographs were handed to me over time, and there is no record of who took them. I, likewise, would hope they will approve of their use as they bring life to the text.

With special thanks to the BOAC Archives, Robert Harding Photography, and to family and friends for their positive thoughts and encouragement at all times. To my daughter, Shelly Stone, who has assisted with everything from design to editing to marketing, and who unfailingly and tirelessly worked long hours so that this memoir would finally be completed. Without her help, it might still be floating in air.

CONTENTS

A Life of Travel
Love and Tragedy

Referring to my old BOAC notebooks, diaries, newspaper articles, press reviews, publicity reports, letters, photographs, and, most importantly, my Flight Logbook, I have endeavoured to paint a picture of past personal memories.

Trustfully, this memoir will preserve the names of the people mentioned in these pages who are no longer with us, and the roles they played on the personal and public stages of our shared history. I hope that those of them who are still alive will feel only the privilege and none of the pain of being remembered.

IN LOVING MEMORY OF THOSE WHO
LOST THEIR LIVES ON FLIGHT 781
OFF THE ISLAND OF ELBA, 10TH
JANUARY 1954

Dorothy Beecher Baker
James Macdonald Bunyan, Alice Bunyan, and Alison Bunyan
William John Bury
Bernard Butler
Evelyn Jean Clarke
John Baptiste Crilly and Brenda Francis Crilly
Anthony Crisp
Elizabeth Fairbrother
Ruth Elizabeth Geldard, Carol Elizabeth Geldard, and Michael
Ian Geldard
Alan Gibson
Francis Harold Greenhough
John Patrick Hill
Marshall Joseph Israel
Rachel Joseph Khouri and Nancy Joseph Khouri
Donald Thomas Montague Leaver
Charles Alexander Livingstone
Francis Charles Macdonald
Ella Sheila Daphne McLachlan
Luke Patrick McMahon
Thomas James Hallanan Moore

Samir Naaman
John Yescombe Ramsde
Frank Leonard Saunders
Howard Edward Schuchman
Robert Sawyer Snelling
John Steel
Chester Wilmot, Vladimir Wolfson
Leila Husain Yateem

IN LOVING MEMORY OF THOSE WHO WERE NOT
RECOVERED
Dorothy Beecher Baker
James Macdonald Bunyan, Alice Bunyan, and Alison Bunyan
William John Bury
Anthony Crisp
Elizabeth Fairbrother
Carol Elizabeth Geldard and Michael Ian Geldard
Alan Gibson
Francis Harold Greenhough
John Patrick Hill
Marshall Joseph Israel
Ella Sheila Daphne MacLachlan
Thomas James Hallanan Moore
Samir Naaman
Frank Leonard Saunders
Howard Edward Schuchmann
Chester Wilmot

Inscribed on the memorial are the words,
Nothing is here for tears,
No weakness, no contempt,
Dispraise or blame.
Nothing but Will and Faith.

John Milton (1608–1674)

The Memorial

FORGET ME NOT

At Porto Azzurro, under the Italian sun,
My friends, my crew from Comet 1.
The flowers faded, the memories clear,
A memorial for ones so dear.
10 steps to the chapel, peaceful, cold,
The iron gates, solid, old.
Rusty lock, rusty chain,
From elements of heat and rain.
Names without faces, clear, precise,
Their dying a tragic sacrifice.
Cypress trees like soldiers stand,
On guard in this foreign land.
Marble crypt, silent sound,
Tears for those never found.
Sad memories only left for me.
January '54 - BOAC.

Lorraine Gray

VISIT TO ELBA, 2009

On a sunny September morning as my flight left Bristol airport in the United Kingdom and soared across the sea, I was excited and a little sad, glimpsing the jagged blue-green coastline. Two hours later, touching down at Galileo Airport in Pisa, I spotted my Italian friend Donatella waiting for me. We were thrilled to meet again and had so much to say to each other. Donatella, a gorgeous, high-cheekboned blonde in her mid-40s, was an interpreter when we first met three years before in Dubai in 2006.

A quick hug and we had to make a hasty dash so that I wouldn't miss the train for Piombino. Traveling via Livorno with its medieval forts and towers we changed onto the shuttle train to the Port of Piombino, where I would catch the ferry to Elba. On the train, chugging along the scenic coastal route, we chatted about our time in Dubai, riding camels in the desert, trying to smoke a hookah pipe and enjoying a sumptuous dinner-and-show with local music and a mesmerizing belly dance performance at a Bedouin retreat. When I mentioned that I hadn't reserved any accommodation on the island, Donatella immediately phoned several hotels in Porto Azzurro and reserved a room at a small family-run hotel, the Baia Blu. I guess she felt responsible for me as I was unfamiliar with Italian protocol and I was an English-speaking traveler.

The journey passed quickly and Piombino soon appeared on the horizon. When we arrived, Donatella helped me with my bags. I had to walk to the waterfront and the main ferry terminal to Elba to board the Moby Line boat for Portoferraio. I was grateful for her help, and once she had me organized on board, she had to leave due to commitments at home in Genoa.

Moby Ferry

I SAT on the rear deck of the boat watching the ship's wake making frothy white waves in the ocean. With the sunlight shimmering on the water, the sea seemed inviting. It was a sweltering day, and as the sun unceasingly beat down on the deck, with a cool breeze that broke up the heat. The short 12-mile voyage took an hour. Still, it was pleasurable feeling the warmth of the sun and watching the wavelets and ripples from the back of the boat. A single squawking gull flew overhead looking for titbits tossed into the water from the culinary area of our vessel. Unwanted morsels of food were also thrown overboard by an animal-loving traveler.

While the surrounding passengers were immersed in their

paperbacks and card games, my thoughts wandered to Tuscany's history as I watched the main coastline vanishing from sight.

Lying off the coast of Tuscany, Elba appeared, the fish-shaped island where Napoleon Bonaparte was banished in 1814 and acted as sovereign from 1814 to 1815. He kept a town residence, the Palazzina dei Mulini, and a country home, the villa S Martino, which I hoped to visit given the chance during my brief stay on the island.

Once the boat finally docked at Portoferraio, on Elba, I ambled down the gangway pulling my small suitcase on wheels. Then walked from the ferry to an area of small untidy shops, to locate the departure zone from where a bus would leave for Porto Azzurro.

Scanning around, I discovered a ticket depot, and from where the buses leave to various places on the island. Standing on the pavement outside the office, I heard a few words of broken English spoken by three young travelers who turned out to be Norwegian. They told me they were also heading for Porto Azzurro before continuing their excursion of the island. The numbers displayed on the outside of the coaches, indicating the terminals, could be misleading. So I remained close to my new friends for their help. While waiting for the bus, I browsed around the few tourist shops. Then, to pass the time I bought a local newspaper and sat on a bench trying to read it and understand the contents of daily events.

No one seem to speak English, so it was essential to have a smattering of Italian intermingled with body language. The Italians are warm, lovely people, and I got help when needed.

Our coach finally arrived at three o'clock. Once it was packed to capacity with tourists and residents, we left Portoferraio. We continued along narrow gravel roads, stopping at various villages and towns around the island. There were precarious drops in the mountainous areas. The long drive took us through a picturesque countryside with distant farms, lush vineyards and valleys. The old bus creaked and rocked whenever we hit an unexpected bump.

This Tuscan island, surrounded by pine-clad mountains and pristine beaches, looked like a fascinating place. Snaking down a winding road from the top of a hill, we descended into the quaint town of Porto Azzurro. Our bus stopped in the town center outside a restaurant facing the major thoroughfare. In the glittering sunshine, the chatty Norwegian girls smiled and waved as they headed for a youth hostel. Gazing around at various nearby shops and small restaurants, I wondered which direction to take to my lodgings. I walked up and down the street, past a maze of cobbled lanes and alleyways with pink, gold and orange stone houses, trying to find my way to the Baia Blu, where I would stay for two nights. I noticed several young Italian men gathered beside a nearby fruit stall. They were talking, while flailing their arms and gesticulating. I approached them, asking about the whereabouts of the Baia Blu. They didn't understand English, but with my very basic Italian, they pointed me in the direction of the shoreline.

Porto Azzuro

I crossed the palm-dotted road, down towards the glimmer of sea, and rounding the corner, walked along the waterfront. Several people were in diving suits, reveling in the surf in the late-afternoon sun. The air was heavy with the smell of the nearby ocean, sun lotion and whiffs of garlic and rosemary from the beachfront trattorias, filled with locals enjoying their early-evening gatherings. At the end of the main road, I spotted the modest family hotel. Inside, my host and hostess greeted me warmly and introduced themselves as Luigi and Antonella. Although they spoke little English and I couldn't speak Italian, we somehow understood one another.

Antonella led me up a flight of stairs to a sea-facing room, sparkling-clean with a private bathroom. She opened wooden trellis doors leading onto a small balcony, overlooking a strip of sandy beach with the sea lapping on the shore. Discarded water skis lay on the sand next to a wooden beach hut. Several hundred yards along the coast was a marina filled with boats, ranging from sea-faring cruisers to smaller yachts, reminiscent of Monaco. This pretty Portoferraio harbor seemed to be for the affluent. The view was outstanding, although traffic frequently passed because of the hotel being on the major thoroughfare through the town. But the noise didn't bother me.

Antonella gave me a leaflet with local attractions and left me to settle in. Hunger pangs being a driving force having not eaten all day, I decided to explore the waterfront to find a place to eat. My hostess had suggested a charming old-world trattoria called Delfino Verde, along the glittering blue harbor.

Tables overlooking the water

I sat down at a small table where I could gaze at the colorful seascape and watch fishing boats arrive after a day's catch. The waiter appeared, handing me a lengthy menu, and told me his name was Enrico. He spoke little English. Anyway I ordered spaghetti with clam chowder sauce and an ice-cold beer. When the meal arrived, it was most unusual. I expected to consume large, lumpy clams, but they embedded the spaghetti with tiny shells containing minute clams. The chowder was tasty and satisfied my hunger.

When Enrico returned with coffee, he asked why a woman on her own was visiting Elba. I explained I had come to find the cemetery to view a memorial of an aircraft that crashed off Elba on the 10th of January 1954. He knew exactly what I was talking about and later discovered most of the islanders were familiar with the shrine.

I asked him if he knew of anyone who might still be alive who would have seen the plane fall from the sky into the sea. After a great deal of thought, he told me there was someone still alive who was a small boy at the time of the crash. He was fishing with his father when the plane dropped out of the sky into the Tyrrhenian Sea, between Elba and Monte Cristo. Enrico phoned various people to find out if he could trace the person and arrange a meeting for the following evening. I could talk to the gentleman with my new friend Enrico acting as my interpreter.

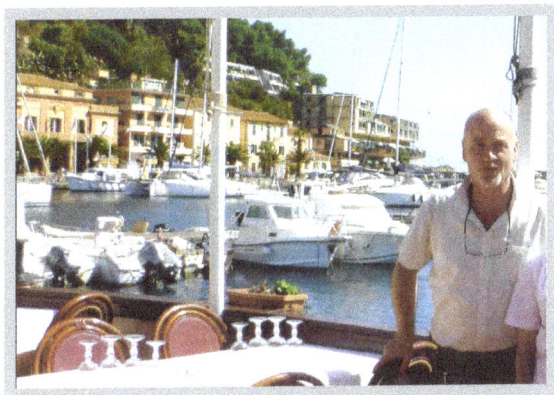

Enrico, my interpreter

After dinner, I strolled back to my hotel along the Grande Lungomare while gazing at the pale-pink glow of the sun setting over the ocean and the distant hills. In my room, after a refreshing shower, I fell into bed, exhausted, surrendering to the noisy traffic outside that acted like a sedative, as I drifted into a deep sleep.

THE CEMETERY

T he following morning, feeling refreshed after a restful night and having breakfast, I left the Baia Blu with a full day ahead. I passed boutique hotels, ice-cream parlours and packed café terraces along the main road, out-of-town, heading for the cemetery. Red and orange bougainvillaea cascading from the balconies of closely packed dwellings added brilliance to the ochre buildings, which were accessed by narrow streets, alleys and flights of steps.

I wandered along the stony, unsurfaced road, with acres of luminous green fields and vineyards on either side. At one point two large stone-pillared entrance gates heralded a long, curving driveway to a small farmhouse at the bottom of an enormous hill. A massive cross stood on top of the hill with a backdrop of vivid blue sky adding to this ethereal scene.

Entrance gates to the farm

The cross on the top of the hill

AFTER A WHILE, I reached a remote area close to the cemetery. Several acres of land were surrounded by imposing 15-foot-high white walls. A cyclist stopped close by to quench his thirst from his water bottle as I stood gazing at the structure before me. The friendly young cyclist was an island resident and informed me that the Porto Azzurro Cimitero was inside the high enclosure. This burial ground was the reason for my visit to Elba.

Walking up a flight of broad cemented steps, I came across the necropolis entrance gates barred and corroded. A thick chain padlocked the two trellised gates, covered with rust and grime, that looked as though they hadn't been disturbed for at least five decades. An archway above the top of the entrance supporting a large ornate stone cross caught my eye. Peering through the rusty old bars, I noticed the end of a paved walkway inside led to a small chapel in the middle of the cemetery. Cypress trees reaching high to the heavens lined the concrete path. It was in this little chapel that they lay the remains of those retrieved from the sea in 1954. Villagers arrived from near and far to pray and place flowers around the bodies.

Old entrance to the necropolis

I wondered how I was going to enter this necropolis when I saw a narrow side road which passed an old, arboreous nursery in need of restoration. This potholed gravel road lined with large trees led to tall, black wrought-iron gates at the side of the cemetery walls. It was the point of entry in use today when people departed this world.

On entering this part of the graveyard, its immense size left me awestruck. It seemed to resemble a maze. A solitary caretaker was standing on a ladder, watering the flower baskets attached to each crypt. There wasn't another living soul in the cemetery apart from the two of us. So I approached him. He climbed down the ladder and came to me. Slight and lean, he seemed to be the same height as me, five-foot-three. His gray hair enhanced his narrow, wrinkled face with a silvering moustache. His kind blue eyes looked very tired though they were still alert. And his handsfree wrinkled and worn from years of work in the sun.

Line of crypts

"Scusi, per favore," I said, asking in a few words of Italian mixed with English if he knew about the memorial of the plane that crashed off the island in 1954.

The little man stared at me in blank amazement. He pointed to a basket of flowers hanging on a lower crypt, with a photograph of a woman inset on the tiny door. I shook my head. He carried on watering the flowers, made a sign of the cross on his chest, and planted a kiss from his hand onto the picture, then resumed his mission.

I was alone and had to continue my search along narrow stony-pebbled avenues, some with large family vaults on either side. Doors were open to many of the vaults, and candles flickered above the coffins. But several of the tombs looked foreboding, with dark sealed doors.

Heading down the walkway towards the main rusty gates, where the cypresses lined the route, I passed graves dating back to the 1700s. Turning onto a gravelly path on my right, close to the entrance, I found the large marble memorial dedicated to the passengers and crew. They died when the BOAC De Havilland Comet, G-ALYP, crashed between Elba and Monte Cristo on the 10th of January 1954.

Once I discovered the monument in this peaceful setting, it was pleasing to see that a thick, towering, green hedge surrounded and protected it. And here, in this remote corner of Porto Azzurro, lay seven graves of the 15 recovered bodies. The remaining eight may have been returned to their homeland for burial.

The funeral ceremony at Porto Azzuro on Elba for bodies recovered from G-ALYP

The memorial was extensive. A long erect marble structure stood at the head of the monument, engraved with the names of everyone aboard G-ALYP. In the earth on the left, on a long, flat marble tablet were the names of all those not recovered from the ocean. One of those was Chester Wilmot, a war correspondent who reported for the BBC during the Second World War, who also wrote the acclaimed historical novel *Struggle for Europe*.

Standing alone and gazing at the simplicity of the entire shrine, I was overwhelmed with emotion as it held a profound story. The late-afternoon heat was still intense, and having walked from town, I felt a little weary. I had no desire to leave immediately and sat on a wooden bench beneath the outstretched shady branches of a large tree to rest for a while. I held a long-stemmed red rose bought from a roadside flower stall on my way to the cemetery. Facing the memorial, and once again gazing around at the large outdoor crypt, I read and re-read every name, some so familiar to me. My eyes closed. Thoughts of the past came flooding back, events that happened somewhere, sometime ago.

THE VOYAGE, 1952

M y sister Audrey and I had spent five of the happiest
months of our lives together, from the
commencement of the sea voyage on the ship the
HMHS Gerusalemme. We were sailing from Beira in
Mozambique to Venice, our ultimate destination. We sailed
along the east coast of Africa, calling at interesting ports along
the way. We cruised past Dar es-Salaam and the enchanting
green island of Zanzibar. The next morning, standing on the
deck of the Gerusalemme and leaning over the railings, we
watched our ship drift into Mombasa's Kilindini Harbour, where
we would be anchored for the day. Kilindini has two harbors, one
for commercial ships and the other for dhow traffic. Our travel
mates Eric and Bob joined us as we watched the distant
buildings coming closer. They were sadly disembarking here to
make their way back to Entebbe in Uganda as their vacation was
nearing the end. Prior to departing and being familiar with the
Kenyan coast, Eric and Bob suggested taking Audrey and me to
Nyali Beach, a small strip of sugary-white sand, and colorful
coral reef that practically disappears at high tide, as a final
interlude. This trip all started one night when my parents were

having dinner with friends and they decided to send me and Audrey overseas. Audrey, who was three years younger, had just finished school. They knew a friend Shirley James in London who owned a well-known travel agency, so he organised the entire excursion for us.

First on the agenda, we caught a taxi to the most well-known building on the island, Fort Jesus, a 16th century structure, designed and built by Giovanni Battista Cairti, a Milanese architect, between 1593 and 1596. Eric mentioned that there was a great deal more history connected to this fort, now a prison. A notice outside the entrance to the fort read: *Mombasa Prison, No admission except on business; Not open to tourists; Smoking prohibited; BY ORDER.* We could visit certain areas of the thick, battered walls. And then we had our pictures taken standing next to an old cannon used in the skirmishes of the 1600s.

My sister Audrey

Afterward we drove to Nyali Beach Hotel, with the beautiful beach beyond it. Audrey, Bob, Eric and I sat on the white sandy shore for a while chatting and feeling contented. The palms were waving in the balmy breeze and a clear blue sky stretched across the heavens. The sea was calm, and in the distance a fleet of high-pooped dhows with patchwork sails were drifting towards the Old Harbor.

Digging our toes into the sand, Eric looked at me, smiled and for some unknown reason said, "People in this part of the world are sensational artisans. Everyone is very hospitable in East Africa. People drive 50 miles for a drink with a friend, on roads varying from smooth ribbons of tarmac to rutted and flooded tracks."

The afternoon quickly disappeared. Standing up and dusting the sand from his knee-length khaki shorts, Eric said, "The time has flown. It's four o'clock, so we must take you back to the ship. Otherwise it will leave without you."

We were sorry to part from our friends who were always cheerful and wonderful companions. So we all returned to Nyali Beach Hotel bar for a quick drink together.

"I hope we meet again someday," said Eric.

"And thanks for a most enjoyable time," Bob added.

I watched my pretty blue-eyed little sister Audrey, her windblown blonde curls flowing around her happy face, returning the salutations. I did the same, as we were hugely appreciative of their companionship and kindness. I wanted to stay in touch with our new friends and asked, "May I contact you if I ever come to Uganda?"

"Please do," they both said simultaneously.

We all laughed. Everything seemed so easy and pleasurable in our young, carefree lives.

Dusk was falling and our chums were eager to start their journey to Entebbe. They would be traveling through the night and sleeping on the roadside in the car they hired for the trip. They had to collect their vehicle from an agency close to the harbor. We hugged and said our goodbyes. Tears ran down our cheeks as we waved our red-headed veterinary officer and dark-haired advocate out of sight. So we caught a taxi back to the port to board the Gerusalemme.

Back on the ship, listening to my rumbling stomach, I made my way to the cabin where I bathed and changed for dinner. Once again, anchors were aweigh and the ship cut through the turbulent waves as we enjoyed another four-course meal.

On New Year's Day 1952, we arrived at Mogadishu. Disembarkation was forbidden, as our vessel was remaining just long enough to take on or transfer passengers to the Somali shore. Very few of the tourist-class Italians left the boat as they were bound for Italy. An Italian officer friend joined me at the ship's railings, leaning over to watch the passengers leaving. Passenger boat's anchored between half a mile and one-and-a-half miles offshore as the Mogadishu harbor was only suitable for dhows and light steamers. Canvas baskets transfer the disembarking passengers from the ship to a small landing craft which ferries them to shore. The transportation procedure fascinated me. I mentioned this to my friend, who'd seen it many times before.

"Tell me something about Mogadishu," I said to my companion.

"Mogadishu is the capital of Italian Somaliland founded over one thousand years ago by Arab colonists. It was known to Marco Polo and many others in his time," he answered and handed me the binoculars strung around his neck. "Look through these across the water. The only prominent building is the Octagonal Tower as it's on Mogadishu's waterfront. It is evident in the Arab architecture vaguely seen."

Handing his binoculars back, I smiled. "You seem very knowledgeable about all the places we've passed."

"I ought to be. I've traveled the route many times," he said in jest, laughing.

The Gerusalemme set sail from Mogadishu in calm waters. New Year's night, after dinner, 12 passengers on board, together with my sister, went onto the ship's bridge to look at the calm moonlit sea. The women in red lipstick and flowy dresses and the men in elegant suits and ties, with cigarettes in hand, were sipping champagne and cocktails against a warm ocean breeze. A golden road of moonbeams divided the smooth dark water. The moon was full, and the African sky glittered with stars. The air was warm and still. The Italian crew began singing romantic songs in melodious voices from the decks below. They also sang popular arias and Christmas carols. We all remained here until their sounds faded into the night and only then retired to our various cabins. The following day at sunset, we rounded Cape

Guardafui in Italian Somaliland, referred to as the Horn of Africa.

On circuiting the horn, our vessel left behind the Indian Ocean and branched out across the Gulf of Aden. In the morning we docked in the Port of Aden. The calm harbor waters before us looked gray and dirty. And a panorama of fishing boats seemed dejected beneath the sky. As we were remaining here until late afternoon, the transit passengers were eager to get off and explore the town. I wanted to see the ancient architecture and the Tawila Tanks, also known as Queen of Sheba Tanks, in the Tawila Valley. This impressive system of tanks used to store drinking water as long as 2,000 years ago was one of the greatest historical engineering achievements in Arabia.

A motor launch showed up to take us ashore. Then we walked from the quayside to the main street of New Aden and found the buildings modernistic in design. There was no touch of the oriental architecture, though the smell of incense drifted from almost every shop door.

Vendors screamed at us to "come in" and buy their wares. They recognized a tourist when they saw one. We wandered in and out of shops only to discover that the town contained many people of the Red Sea managing the stores, and general items were affordable.

Arriving in Old Aden midday, Audrey and I went for a hearty Arabic lunch lasting a few hours, and learned to smoke a hookah pipe. After the meal we continued to the Tawila Tanks before returning to our ship late afternoon. Once back on the Gerusalemme, bumboats carrying provisions for sale surrounded the vessel. Their peddlers hoping to bargain with the passengers prior to sailing. They were stocked high with carpets, silk scarves, jewelry and many other items to tempt the last-minute shoppers. The bumboat men knew the last word in sales talk as they hawked us their wares, and it was at this moment that I weakened. From this port of call before entering the Red Sea, I bought several silk scarves as souvenirs for the folks back home, including one for myself. All the remaining travelers stood on deck watching the fading sun disappear on the horizon. We waved to the owners of the small boats as the Gerusalemme sailed out of Aden Harbour.

It was still very early when we arrived in Suez. Audrey and I slipped into our dressing gowns and went on deck, thinking we would be the only solitary figures that had risen at the forsaken hour. To our surprise, several passengers had already beaten us to it, loitering about on deck as the waking sun rose over the far-off hills. Perfect time for photographs.

Anchored in the bay a couple of destroyers looked very warlike silhouetted against the darkish, shadowy expanse of the distant hills. We were currently waiting to take our turn in the queue to sail through the Bitter Lakes and the Suez Canal. During this interlude, my sister and I returned to the cabin and got dressed for the day before the ship's engines started again.

Midday on the Red Sea was like sailing through a furnace, not a breath of air and the sun beating down on the decks. We as passengers escaped the noonday heat by sitting in an airconditioned lounge and drinking iced sodas. We also cooled ourselves in the pool as we coasted through the Red Sea's calming waters. From time to time, dashing to the rails of the boat to see if any sharks were visible.

We knew that our vessel had anchored in the Suez Port when the vibrating engines of the Gerusalemme were no longer felt. Suez was the first port of call in Egyptian territory. As antiBritish riots took place between Suez and Port Said cameras were forbidden. Flaunting the rules, we were determined to snap a few pictures of Egypt. So we would have photographs to look back on of this happy voyage and all the countries passed through.

Before making its way across the Mediterranean Sea to Italy, our vessel moored at Alexandria. Finally, docking in Venice, where we got off excitedly. My sister and I would spend a week in this breathtakingly beautiful city and had made reservations at the Hotel Danieli, a magnificent 14th-century palazzo on the Grand Canal.

Our first day was fantastic. We viewed the island from a gondola with a gondolier wearing a stripy blue t-shirt and hat to match, serenading us with local music. In January the Venetian fog makes the city quite mystifying and temperatures rarely drop. The festive spirit of Venice is always on display, including the charming antics of the local pigeons in St Mark's Square.

On Sunday Audrey and I rose early in time for the church service at St Mark's Basilica. Its interior is renowned for its opulent design and gilded mosaics and is called the Church of Gold. Before leaving Venice, we visited the Islands of Murano for glass-making and Burano for lace-making and bought several pieces of remarkable glassware for family and friends back home.

Our next stop was Zermatt, in Switzerland, where the ski season was in full swing. We stayed at the luxurious Mont Cervin Hotel for two weeks. The day after our arrival we shopped in the village stores, spending a reasonable amount of money on glamorous ski outfits, and signed up for the ski school run by Gottlieb Perren. He and his brother Bernhard were training for the Olympics taking place in Norway, apart from assisting novices on the snowy inclines. After a great deal of hard work, practising different basic movements on the slopes, we were both given a simple medal for our achievements. The snowplough, parallel skiing, stem Christiana, and turning on skis were some of the things we had to learn. Following a hard day on the slopes, everyone gathered at the hotel bar with other residents, drinking *glühwien*, the German equivalent of spicy mulled wine, and dancing until midnight.

One evening, an après-ski fondue party took place in a rustic cabin in the mountains. The ski instructors arranged it for all the new skiers. Fondue parties would later become the rage in the 60s and 70s. Everyone walked up to the chalet, carrying their skis. A table was set with an enormous communal fondue pot. Beside a warming fire, we shared the pot of hot, melting cheese and warm bread. Then, after a cozy night out on the tiles, we all donned our skis and skied down the slopes back into the village beneath a sky full of stars.

Leaving Zermatt, our travels took us across Europe by locomotive to England.

ARRIVING IN LONDON

W e landed in England on February 7th 1952, in time for the funeral of King George VI. He died the day before our arrival, and the funeral took place on the 15th. King George was the last Emperor of India and the first head of the Commonwealth. Our maternal grandmother's family had deep roots in England, going back as far as the 17th-century. After the First World War, when my grandmother lost her son and husband, she and her brothers immigrated to Northern Rhodesia (now Zambia) and started farming and subsequently created many successful businesses. We were sent by train to South Africa to be educated at St Annes College, a private girls school. A number of years later, some of the family would return to the UK and some went to the USA.

After a fortnight in England, Audrey and I flew to Norway to watch the Winter Olympic Games in Oslo. The ski jumping at Holmenkollbakken was the highlight of the journey for us. Here we watched the jumpers soar through the air like birds in flight and then land gracefully on the snow. In total we spent four awe-inspiring days enjoying the winter sports.

The 23rd of February, on the voyage back to England, we strolled the deck of the boat in the cold misty weather. As

Audrey and I stared out at the glimmer of foggy coastline in the far-off distance, a dark-haired young man wandered over and starting chatting. Wearing a thick overcoat, covering his ski clothes, his right arm cast in a sling, he introduced himself as Rupert de Larrinaga, an Olympic skier for the British team. Rupert was returning home two days before the games ended as he had an accident while streaking down the ski slopes during a practice run. His right arm had been broken in several places. We had a wonderful time talking about the exciting events that took place. Before leaving the boat, addresses and phone numbers were exchanged with plans to meet in London.

Over the following months back in England, Audrey spent time in the north with friends. She visited the sights of London on the back of a motorbike owned and driven by her friend Ken from South Africa. They toured France venturing to all the places of interest and travelled as far afield as the German border.

Three-and-half months of glorious globe-trotting had come to an end and I realized that my desire to return to my hometown in Rhodesia was still far off. What type of job could one do to keep on the move? My brain kept turning like a merry-go-round.

There was the unavoidable problem of accommodation. If I remained in England, I could not continue living in hotels. Besides, it would be inconvenient living in one room, apart from not having a feeling of comfort and homeliness, particularly after a long day in an office. It would be so much more pleasant sharing a flat with a friend or several friends.

My sister and I had a reservation at the Earl's Court AA Hotel. It was mid-winter and the place was bleak. We hoped that our room would give us a feeling of warmth. The hotel porter escorted us to a spacious room on the first floor with two huge sash windows overlooking the high street. He then left us with the words, "If there's anything you need, please let us know."

A slight chilly breeze drifted in through the sides and bottom of the rickety, battered frames of the windows, which appeared to have seen better days. The gloomy wet street and the sky above added to the general gray appearance of everything outside. Cars looked gray and the sooty old buildings looked gray. Even the few birds circling the rooftops looked as though they had flown through gray smoke emerging from chimneys. The gloom and lack of color was startling. I closed the heavy red curtains, hoping to keep out the cold draught. The fireplace in the room appeared to be a gas fixture. Raised in a small town in northern Rhodesia, where the climate was consistent and mostly warm, we had never seen gas heaters. There was no need for fires unless one had a log fire on a sunny, cold winter's day.

Trying to figure out how to warm up the room, I studied the sign showed by an arrow close to the heater, which seemed simple enough: *Turn knob in this direction to start.*

I followed the instructions, but after about 30 minutes there was no change in the atmosphere, merely an unpleasant smell. The room was freezing and we could see our breath. Audrey and I, fully dressed in multiple sweaters, curled up underneath the quilts on our beds, shivering and trying to warm up. It never occurred to us to call the janitor for help.

"Strange," I said, "I've turned on the heater as instructed, but there isn't even a spark flickering. The smell in the room is overpowering." I leaped off the bed, ran to the window and drew back the curtains, opening the windows to clear the air. We lay on our beds buried in quilts and blankets until six o'clock in the evening. Our friend Rupert, the dashing skier with a broken arm whom we met on the boat returning from the Oslo Winter Olympics, arrived to take us out to an elegant restaurant for dinner. When I opened the door, the smell of fumes overcame him as he stepped inside.

"Rupert, I tried to light the fire, but there was no flame, even though I switched on the current. What do I do to heat the room?" I asked.

He look concerned and, without saying anything, rushed to

the fireplace to turn off the gas heater. "You know there might've been an explosion if anyone lit a match in your room," he explained. Our new friend saved our lives as we would've gone to bed that night, having slept with windows closed to prevent the bitter cold from blowing into the room, and fumes emanating from the fireplace.

My sister and I intended to stay at the Earl's Court Hotel for a week. One evening a friend unexpectedly phoned, inviting me to a performance of the opera *Turandot* at the Royal Opera House in Covent Garden. Events in my life always seem to happen spontaneously.

I dressed up and dashed off to the tube station to board the Piccadilly Line. While waiting outside the Opera House for my companion and the doors to open, by chance, I saw Paddy Pendray, waiting in line. She was an old chum from school, now studying physiotherapy at St Thomas' Hospital in London. In fact, during our time at St Anne's College in South Africa, we were in the same class. There were many stories to relate from our last day in college until this moment. During our conversation, I told her I was keen to stay in England, but having trouble finding a suitable place to live.

"Come and join our group at 94A Cromwell Road," she said without hesitation. "We'd love to have you. There are another three girls in the flat. Roz, who is training to be a doctor, Veronica who works for an embassy in the West End, myself, and Val, who's on the road to becoming an opera singer."

Thrilled, I thought it was fate and wasted no time accepting her invitation. "Thank you, Paddy. That's wonderful! When can I move in?"

"Tomorrow... or whenever it's convenient. Must go or we'll be late for the opening. See you anytime it suits you." She quickly jotted down her number.

I thanked her again, and after meeting my friend in the portico, we rushed to our independent boxes.

~

THE FOLLOWING morning Audrey and I left the hotel and went our separate ways. She went north to stay with relatives in Bolton and I hailed a cab to drop me with my trunk and suitcase at the flat in Cromwell Road in Chelsea. Fortunately, the driver could stop right outside the house.

I got out of the cab, walking up to the front door, rang the bell, and waited for several minutes before Paddy arrived to welcome me. It was her day off, so she was at home swatting for an exam, which was taking place at the end of the week.

The taxi driver helped me carry my trunk up the flight of stairs to the first floor, followed by my suitcase. I handed him a reasonable tip, for which he thanked me and left.

The flat was spacious with a comfortable lounge, three good-sized bedrooms and a small bedroom. Roz, the doctor trainee, occupied a smaller room of her own for studying at all hours. Veronica, the opera singer, shared a room, and Paddy and I halved the second-largest bedroom. They kept the smallest room for visitors. A spacious lounge with a bay window faced an unused, overgrown garden at the back, with several elms and oaks that appeared to be a haven for a variety of birds. Sizeable windows at the front overlooked Cromwell Road, giving us a panoramic view of activity in the area.

After settling into my new home, the next step was to find a suitable job. But what? At times I felt the bright, dazzling parts of my life had already happened. So, what could I do to continue traveling and working? Would it be possible to become a secretary to an itinerant business executive, a fashion model or even a movie star? These sought-after positions were rarified and like reaching for the moon.

One morning I went into my bank in South Kensington to withdraw money and saw a girl in uniform leaning over the counter. She was busy talking to the cashier. Possibly an airline employed her as she looked so chic in her sharp navy suit sculpted to her body, her dark hair swept back beneath a jaunty air force-style cap.

I stared at her, thinking, *an airline stewardess, could this be possible?* I had heard it was a coveted job by people all over the world, both near and far. I knew competition would be sky-high, but there would be no harm in trying, as a whirling, nomadic life would suit me better than anything else. After celebrating the best birthday of my life with Audrey whizzing around the coast of Africa to Europe, I realized I was a wild-hearted wanderer, an adventuress, with an insatiable curiosity. All I wanted was a different and daring life and to see the world.

When the girl turned to leave the bank, I gasped as I recognised her. She looked stunned and let out an excited hello when she caught my gaze. We couldn't believe that after leaving our college in South Africa, the two of us should meet in a bank in west London. Here before me was my old friend Anne Ryder from school. Being two years my senior, she graduated before me and left for England soon afterwards. Anne invited me to visit her at her north London flat, so we could catch up on everything since parting ways a couple of years before. During our brief encounter, she suggested I join the ranks of the chosen few and apply to BOAC. The British Overseas Airways Corporation was the state-owned airline.

"Anne, it seems to be very difficult to get work in an airline. I've heard from general remarks that the job is desired by many people."

"When you come and see me, I'll tell you more about it and then you can decide." She scribbled her address and telephone number on a card, handing it to me. "Give me a call. Perhaps we can meet next week before my trip to New York. See you soon," she said and walked away.

～

AFTER VISITING Anne the following week, I sat captivated as she told me about her dizzying life of waking in Paris for breakfast and having dinner in New York in the same day. She suggested I apply for a position as a flight stewardess. Instantly my mind was made up, and I was going to be a stewardess. Out of a spirit of adventure, I phoned the British Overseas Airways Corporation for an application. When it arrived in the post several days later, overwhelmed with excitement, I immediately filled in the questionnaire. That same day I spent a few hours researching the form and answering the questions. After reviewing the document, I sealed it in its official envelope and hurried to the South Kensington post office. Putting it into the special-delivery box to ensure that it was in the mail.

It was a freezing, gray, rainy afternoon as I rushed back to the flat along the paved sidewalk. As I pushed past the bustling pedestrians, some of them holding colorful umbrellas, my thoughts raced. *How long would it take to receive a reply? Will they accept me? Am I the right height and weight? Would my qualifications meet the requirements?* A swirl of questions circled around my brain.

Back at the flat, I collapsed onto the settee in the lounge, worn out from the day's activities. Paddy had just returned from training a few moments before. Seeing me soaked and shivering from the icy rain, she made me a cup of tea to warm me up and soothe my nerves. Handing me the tea, she presented a plate of ginger biscuits, and asked me about my day.

"I've applied for the stewardess job I told you about, and hope it will come to fruition."

"Relax, you can't hasten anything."

By now, it was latish afternoon/early evening, and the other girls all shuffled in, gathering in the lounge to discuss recent events, and their lives, over tea. The conversation turned to films currently showing.

"Has anyone seen *Where No Vultures Fly?*" Veronica asked.

"Oooh yes," some of the others shrieked in unison.

"Isn't Anthony Steel gorgeous? He's so devastatingly handsome," said Paddy.

He was every girl's idol. His boyish good looks of dark, wavy hair added to the beauty of his suntanned face with deep-set blue eyes and perfect white teeth and made him, alongside Dirk Bogarde, one of the biggest British film stars in the 1950s.

"Very articulate and a divine voice," interrupted Val.

Veronica then confided that her family knew him when he was in the Gurkha Paratroopers in Gulmarg during the war, as they were living in India then.

After everyone shared their views, since Veronica knew him, I said I would write and invite Anthony Steel to a cheese-and-wine party. "You never know, he may surprise us."

While we were chatting about the upcoming Oxford Ball and Boat Race, I grabbed a notepad impulsively. Caught up in their enthusiasm, I penned a letter in my best handwriting and read it out to the girls for their approval. Then we wandered into the kitchen to prepare supper. There was a knock on the front door and three Oxford students burst into the flat. They were old friends from Michaelhouse College, a boy's school in South Africa, who were now at university in England. Michaelhouse is the brother school to St Anne's college, our old school.

Russell waved a piece of paper and called out, "Girls, girls, we invited Elizabeth Taylor to the Oxford ball. But she's refused because she has other commitments. We're so disappointed." An old friend of Paddy's, Russell was a six-foot-tall, well-built guy with thick blonde hair and a strong squarish face. His heavy dark-framed glasses emphasized his enormous blue eyes and gave him the appearance of an aristocratic newspaper reporter.

We countered their exuberance by letting out a massive sigh like a bunch of drama queens. "Ahh, well, we can better that. We intend to invite Anthony Steel to the flat for a small party," I said, waving the note around. "Envious?"

"Absolutely not," Russell scoffed, laughing. "You don't stand a chance."

The other two boys laughed too.

"We'll see," I said, grinning, and with a giddy mix of joy and nervous energy, dashed out of the flat into the rain to post the letter in a red postbox at the end of the road.

Then we spent the rest of the evening playing Monopoly, with all competing for West End properties. Our friends eventually left at midnight, having taken over most of the valuable areas on the board.

~

A WEEK LATER, as I was happily settling into my London pad, Audrey phoned me to say she was returning to Africa and had reserved a passage on the Edinburgh Castle. She booked it through a very influential friend of the family, who owned a travel business in Sloane Square. The same friend who'd originally organized our trip from Africa. He had contacts all around the world. The ship was sailing the following weekend. I told her I would meet her at Waterloo station three hours before the boat-train departed for the Southampton docks. We would have time to catch up since our parting several weeks ago.

"I'm not going back with you," I told her over the phone. "I've applied to BOAC to become a stewardess and am hoping they'll call me for an interview."

My sweet and anxious sister was surprised. "It'll disappoint the family. They're expecting us home together."

"This is the chance of a lifetime and I must take it. If it fails, then I'll return home on the next ship."

"Well, I'm sure London will be fun and as long as you're happy."

I knew my parents had a different sort of life in mind for Audrey and I, hoping we'd return home to work in the family business and get married.

After the call ended, I rushed downstairs to collect the mail. Among the letters I picked up there was an envelope addressed to me in writing that I didn't recognise, so I hastily opened it. It was a note from Anthony Steel in reply to my letter. A photograph dropped out onto the floor. At the top of his letter, on the left-hand side, he had typed his phone number. I screamed with excitement. After a very brief respite, steeling my nerves, I picked up the phone and dialled Kensington 3191. At the other end, a magnetic voice answered. The call overwhelmed me for a few seconds and I mumbled clumsily, "Um, hello, am I speaking to Anthony Steel... the actor?"

"Yes," he replied in that unmistakable voice.

"It's Elaine speaking, and thank you for your letter. My friend Veronica, who shares our flat knew you, or should I say, her parents knew you, when you lived in India and served in the Gurkha Paratrooper Division, in Gulmarg. As you suggested, I'm calling to make arrangements for you to pop in for a drink when you have a few spare moments."

"Thank you so much, but as I mentioned in my letter, I'm going into St Thomas' Hospital for a minor knee operation next week. I fell off a horse in Kenya while we were filming *Where No Vultures Fly*."

"Oh I'm sorry... and I wish you well. Could we drop into the hospital and visit you after the operation. I'm sure Veronica will be thrilled to see you again."

"It will be a pleasure, thank you for calling," he answered, "See you soon."

48 Malvern Court,
S.W. 7.

Tel: Ken. 3191   23rd August, 1952

Dear Elaine,

     Again thank you for your letter which I was
delighted to receive.   It was nice of you and
Veronica, of whom I have vague recollection in
Gulmarg but who at the time was possibly fairly
young.   I would like to drop in at the flat
for a drink sometime and meet you all but I
go into hospital next week for a minor knee
operation which doesn't leave very much time.
Why don't you call me and we will arrange a
more definite time?

               Best wishes to you all,

                    Sincerely,

                    Tony Steel

Miss Elaine Baker,
94A Cromwell Road,
S.W. 7.

Letter from Anthony Steel

Anthony Steel

My knees turned to water. I couldn't believe I'd spoken to this famous movie star. I was over the moon and longing to tell my flatmates about the brief chat with Anthony Steel when they arrived home.

At six o'clock, I anxiously waited for the flat door to open, and have the girls bursting in crying out for a cup of tea. Paddy arrived first. Threw her jacket and handbag onto her bed, then came into the lounge and curled up in one of the large armchairs.

"You'll never guess what happened," I said, on my way to the kitchen to put the kettle on.

During the next 20 minutes, once the rest of the group came home and were comfortably settled, I related the gist of my conversation with Anthony Steel, brandishing his letter in front of them. Astounded and excited, they couldn't believe I had made contact.

"Well, Veronica, we will have to think of a small gift to take to the hospital. Can anyone suggest anything?"

"Sure! I'll give him some of those large, delicious peaches that our barrow boy on the corner sells on his stall," Veronica suggested.

"Sounds good. I saw a book in a little shop close to South Kensington tube called *Variety is the Spice of Life*. I'll give him that," I said. "He has so much variety in his life it should appeal to his better judgement."

Before retiring, we spent the next couple of hours discussing diseases, opera, music and the day's news.

Now, after several months of globetrotting, it was time for my sister to return home to Rhodesia. Friday, the day of her departure, we met at a cafe at Waterloo station, where we caught up on everything. I reminded her I had sent an application to BOAC, but still hadn't heard back.

When our conversation ended, we finished our coffee and walked to the platform from where her train was leaving. Travelers packed the station. Young lovers with clasped hands, people chattering when they would meet again, others hugging and shedding tears. We stood and watched these scenes when the train slowly shunted into the station.

Everyone said their goodbyes and boarded as soon as the carriage doors opened.

The loudspeakers were blaring old sentimental refrains, such as *We'll Meet Again, Now Is the Hour* and *Auf Weidersehen*, which was rather fitting. Partings always felt a little sad. Audrey and I were both teary. I hugged my sister tightly, as if I may never see her again, and waved her off. As the whistle blew for departure, carriage doors slammed closed, the loudspeakers pealed the final tune, *Auld Lang Syne*. The crowds on the platform waved while watching the train pull slowly out of the station until it rounded the bend and was out of sight.

Audrey was a keen snapper and our trip around Africa and Europe had sparked her passion for photography. She ended up opening her own portrait studio in our little hometown of Broken Hill.

~

ON A COLD, rainy Monday evening, as soon as Veronica got home from work and changed, we picked up the gifts for Anthony Steel, walked hastily down the stairs of the flat and out onto the glistening pavements. The lights from the streetlamps shone in the puddles on the damp streets. We caught a train to Westminster tube station to reach the hospital. Leaving the station, people eager to get home after work were rushing along the sidewalks. Some looked harassed or disturbed with furrowed brows, while we were in our element and ecstatic heading to meet a movie star, our film idol.

"Now remember, we mustn't outstay our welcome," I reminded Veronica as we crossed over Westminster Bridge.

"Absolutely, we wouldn't want him to become bored during our first encounter."

The hospital's location opposite Parliament was convenient, so we could see Big Ben across the river with its enormous face and hands.

"But we might not see the clock from his room," I said.

"It's a chance we'll have to take," she replied, on our arrival at the hospital.

We were escorted through several corridors to his private

ward after our arrival at the hospital. The nurse opened the door and announced our entrance. Propped up with pillows, Anthony Steel looked stunningly attractive, square-jawed, with blue sparkly eyes and a brilliant white smile. Sitting up in his comfortable bed, he was wearing smart blue-and-white-striped pyjamas. He held out his hand to Veronica. "How are you, darling?" he said, giving her that familiar gorgeous broad smile.

She shook his hand shyly. "Thank you, I'm well."

My palm was moist from excitement, so I unobtrusively wiped my right hand against my coat and then, introducing myself, shook his hand. The nurse brought in a couple of chairs which she placed at a certain angle, and, amazingly, gave us a full view of Big Ben across the Thames. We kept an eagle eye on one another; only half an hour transmitted from Veronica's brain to mine. Conversation flowed as we talked about Africa and the lifestyle out there, and he talked about about shooting his last film in Kenya. Then he started telling us about his next likely film when Big Ben pealed 8:30pm. This was our cue to go.

We stood up, said how pleased we were to find him looking so well, and hoped to see him soon. As we were about to leave, flashing a dazzling smile at me, he asked, "Have I got your telephone number?"

"You have Veronica's, but you don't have mine."

He had an impish sparkle in his pale-blue eyes, openly flirting with me. "Please let me have it so we can be in touch." Maybe he thought I was desirable and available.

"There's a possibility I may relocate soon, so I'll send it to you once I know," I said teasingly, but meant it. "I have your address."

Veronica and I were ecstatic, giggling all the way to the tube. This was possibly our last encounter with one of the most desired idols of the film world.

MODELING SCHOOL

❦

One afternoon I glimpsed through a glossy leaflet for a London modeling school that had dropped through the post. It included photos of a glam, doe-eyed model gazing out of spidery eyelashes and an article written by her on how-to-be-a-successful-model and all the excitement and exotic travel that it entailed. I never fancied myself glamorous enough for such a lofty career, even though as a teenager I had been stopped in the street a few times and told "you should be a model." I was gray-eyed, high-cheekboned with a leonine mane of ash-blonde hair and a slim, slight figure, but at five-foot-three, didn't think I was tall enough nor stood a chance. Since there was still no news from BOAC, I had already signed up for evening French classes at a local academy, but thought I may as well sign up for modeling classes run by an agency. I made an appointment for the following afternoon.

When I showed up at the agency the next day, a smartly-dressed secretary directed me to the school's manager Miss Stobart. She welcomed me and began plying me with questions: "Why are you interested in being a model? Tell me about your education. Do you exercise daily? How well do you think you communicate with people?"

Feeling a little in awe of her questions, I replied to the best of my ability. At the end of our discussion, she checked my weight and height, and registered me. She made it clear that the agency was highly regarded in the fashion industry and that these doors were crossed by many famous faces.

~

ALL THE NEW models in our class gathered in the basement of a large redbrick building on Sloane Avenue in Chelsea the following week. The classes started at nine o'clock sharp each morning, finishing at two or three in the afternoon. The basement had a T-shaped stage, a sort of catwalk, and mirrors on the surrounding walls. An entrance/exit door for the audience was at the back of the podium. The only gap into the actual model hall itself was the door on stage with a draped curtain falling from the ceiling to the floor, along with a backstage area with dressing rooms and makeup cubicles.

Mornings in London were energizing. I would rise early and have a light breakfast of grapefruit and coffee, and later a fresh salad and vegetables for lunch to keep a clear skin and healthy weight. In the bright morning air after breakfast, I left the Cromwell Road flat and scurried to Sloane Avenue. My flatmates were still in bed as they were starting work an hour later. The night before we had all been to a friend's flat in Soho for a birthday party, but I didn't have a sip of alcohol, so I could look fresh the next day. During the next three weeks, the model school provided us with an intensified course, teaching us to move, stand gracefully, do perfect makeup and wear clothes with style and how to glide across the stage.

During our short mid-morning breaks, two of the girls would dash across the road to a small tea shop to buy a pint of milk for each of us for our coffee and tea. Sweets, chocolates and cakes were out due to them being bad for our skin and it was essential to keep trim. The third and final week of our course arrived. We spent each day perfecting everything that they had taught us for the Passing Out Parade on Friday. Every day was a dress rehearsal day, and the tutor hoped that by now we had developed our own fashion sense. For the big day we had to wear our own clothes and accessories, consisting of a suit, swimwear, lingerie, casual afternoon dress, followed by a cocktail dress or evening gown. Dressing up was all great fun. Several of the girls made their own ensembles and frilly hats to match. They looked smart enough to step into New York's annual Easter Parade. Finally, ten shapeless heads of hair were restyled by a stylist brought to the school. As my hair fell below my shoulders, I had it swept back into a glamorous French twist.

Near the end of three-week course, Miss Stobart gathered us all mid-week and praised us. "Girls, you have worked hard and shown excellent results. While every attempt is made to secure a commitment for you, the agency cannot guarantee jobs. Success depends upon personality, ability and enthusiasm. I wish you the very best of luck for the future."

Friday, the dreaded day, arrived. To appear at the agency revitalized, I hailed a passing taxi as I stood on the pavement. Everyone showed up an hour earlier than usual, so we'd be made up, dressed and in a relaxed mood by the time the guests arrived. The hall filled with people. We were putting the finishing touches to our swimsuits, shorts and everything that goes to make a standout outfit. A general bevy of beauties milling around the dressing room while waiting the curtain call.

We listened to feet shuffling and the buzz of voices sounding like swarms of bees on a hot summer's day. Here we were, gathered together in a Chelsea basement. Fashion editors, famous photographers, reporters and representatives of the prominent wholesale fashion houses filled the space. All looking for unfamiliar faces and talent to whisk into the never-ending whirligig of fashion. Miss Stobart and her secretary sat in the front row.

While waiting in the wings we felt edgy with excitement yet

optimistic, wishing to become top-line models. The music began playing *A Pretty Girl Is Like A Melody*. This was our queue to line up at the side of the stage. The first of our girls had already walked across the stage into the crowded room while we stood in the passage behind waiting our turn. Peeking through the curtain, I noticed cameras flashing. In contrast, other onlookers jotted down notes into their little books. We were all absorbed in watching the participants before us. I noticed one or two of the girls, despite being glamorous, seemed to hesitate as they made their last round of the platform. Wobbling a little on their high heels after a flamboyant twirl and a self-conscious smile, they hastened into the wings to the tune of *Pretty Baby*.

Facing the crowd, I was confident because at the back of my mind I knew that I was only here to bide time while waiting to hear from BOAC. My turn finally arrived, and I drifted onto the stage. After walking sedately down the stairs, I made a flourishing turn in front of the audience. Then, walked back up the stairs onto the platform and, after a sweeping curtsy with gliding movements, retreated behind the curtain. After the last contestant made her appearance, the excitement died down. The continual drone of voices receded as the mass of spectators departed, and amidst all the activity the morning soon passed.

Miss Stobart burst through the door when we were busy in the dressing room, changing back into our street clothes and packing our suitcases. "Girls, girls, you were tremendous," she announced breathlessly. "The photographers are very pleased with the results. Please report to head office at ten o'clock on Saturday morning, when you will receive your various assignments."

"But we have only completed our début," said Laura, questioningly.

"Photographers, the press and showroom clients were most impressed with the outcome and there will be placements for you all."

We left the building in Sloane Avenue overjoyed. As we may

not get back together again apart from on the morrow, we wished each other all good things for the future.

I awoke early on Saturday to be at the agency by ten o'clock. I was feeling tired yet satisfied with the progress I had made during the last three weeks. Not only with the modeling course, but also with my evening French classes. Lying in bed, half-awake and half-asleep, I had almost forgotten about the significant letter I was expecting. I heard the flat door click. Paddy returned after going downstairs to find out whether the postman had left any mail.

"Anything for me?" I called down.

"One from Rhodesia. You're so lucky. I haven't heard from my family for two weeks."

Paddy handed me the letter and, as I tore open the envelope, I said, "Ah well, let's see what's happening in darkest Africa."

She walked out of our bedroom to distribute letters to the other girls. On the whole, we received a fair amount of mail each week from various relatives and friends.

After reading through my letter from home, I suddenly remembered I had to be at the agency by ten. Flying out of bed, I rushed to the bathroom, quickly getting dressed, as there was little time to spare before catching a bus to the studio. My companions went about their business in a relaxed fashion since it was the weekend.

I waited for my turn at the head office for Miss Stobart's interview. Laura, a girl with whom I had become friendly, was sitting beside me in the waiting room.

"I have butterflies in my tummy and I'm feeling rather anxious," she said.

"Don't worry. I'm sure they'll give you an assignment," I answered, attempting to comfort her.

Then I was called into the office, where the manager again recorded my measurements and type of appearance. They offered me a showroom position at a popular London department store, along with a head-and-shoulder photo shoot for the cover of an exclusive magazine.

"Miss Stobart, I would prefer to accept the latter as I've applied to an airline for a stewardess position, but still waiting to hear from them. The showroom modeling is on contract and I would rather not be restricted."

"I understand. Here's the name and address of the magazine. Good luck," she said, smiling and shaking my hand. As I turned away, she announced the name of the next new model.

I swaggered out of the agency and came face to face with another girl with whom I had struck up a strong acquaintance throughout the classes. "Hello Marie, finding out your destiny?"

"Yes," she said, "I'm worried and feeling very indecisive. Don't think I'll be with the agency long. A film producer has proposed to me and I'm thinking about accepting his proposal."

"I'll wait for you in the little coffeeshop on the corner. Hurry along and let me know what you draw out of the hat."

After her interview, we sat in the coffeeshop for well over an hour discussing the past and our forthcoming assignments. Then we boarded our respective buses home, after wishing one another happy times for the future.

THE INTERVIEW

I t was a bright Monday morning, when I tiptoed downstairs, trying not to wake the other girls, rushing out early, to keep my appointment with the magazine photographer. Feeling fresh and excited, I stopped at the bottom, picked up the letters pushed through the slot in the door and skimmed through them. At the bottom of the pile was a pleasant surprise. At last, a letter from the airline had finally arrived after what seemed like an eternity. In reality it was only four long weeks of waiting. I tore open the envelope and read the contents of the letter asking me to come along for an interview on the 9th of July.

Overwhelmed with excitement, I dashed back upstairs to hand the rest of the mail to my flatmates, while telling them my happy news. I folded the precious letter with shaking hands and put it in my handbag. Picking up my small suitcase containing makeup and a change of clothes, I breezed out of 94A, feeling more optimistic than I had a few minutes earlier. This was one of those days when everything felt right in the world.

When I got to the studio, I was immediately ushered into a cubicle to apply makeup and change for my first shoot. The hairdresser added the final touches. One had to choose flattering colors to complement skin tone and enhance the lighting effects of the pictures.

I was a novice at the game, but the photographer was very kind. As he arranged his equipment, he talked a great deal about the varied positions of the head with added facial expressions to complete a picture. He told me to follow his direction on where to look and to keep the poses drifting but alive. His friendly chatter put me at ease as it helped me relax and gain that extra self-confidence in front of the camera.

It took almost an hour to complete the sitting and the photographer seemed happy with the pictures. I thanked him and dashed back to the dressing room to change into my street clothes. Once I gathered all my bits and pieces together, I headed for the agency. When I arrived at the office, most girls had been offered positions and had already left. So, without waiting, the secretary sent me straight in to see Miss Stobart. I told her that I got an interview with the airline the following week. If successful, I wouldn't be returning to continue modelling.

She was full of understanding. Her penetrating eyes looked straight at me. "I wish you all the luck in the world and hope that your desires come to fruition, but if you change your mind, please come back to us."

"Thank you so much. I've enjoyed my time with you and will let you know the result of my interview. Goodbye," I answered, shaking her hand.

～

THE DAY of the interview arrived, and from the time I awoke, I had butterflies in my stomach. The early morning was spent pressing my clothes and polishing my shoes ready for the meeting. I had to make sure I looked smart and impeccable, so I dashed out to the local hairdresser to have my hair shampooed and styled. Afterwards, I bought a couple of fashion magazines from a news vendor at a roadside stall, thinking they would help me unwind and take my mind off the forthcoming event later that afternoon.

Sitting in the lounge with a cup of tea and flicking through magazines, I tried to clear my head and unwind, but couldn't concentrate. The other girls had already left for work early in the morning. Before dressing up for the interview, I relaxed on my bed and started reading *The Cruel Sea* by Nicholas Monsarrat.

A little after midday, I took the bus heading for Brentford and Airways House. When I got there, I saw several other girls in the waiting room, anxiously skimming through magazines. One or two cast a glance at me. I folded my jacket, placing it over the back of my chair, and sat next to an auburn-haired, fair-skinned girl with dark eyes. She was of medium build and very pretty. As we chatted, her warm, outgoing personality put me at ease. It was during this idle chitchat about our lives in London that we introduced ourselves.

"My name is Jo Stanbury. I'm from Glastonbury."

"Elaine Baker, from Northern Rhodesia."

Then the door to the interview room opened and a gentleman called her name.

"Good luck," I whispered. "I hope we meet again."

"Good luck. I'll hold thumbs for you."

When Jo returned from the fated room, she popped her head around the waiting room door and called, "Bye in the meantime."

As I sat in the chair, thinking if I failed the interview, I would backtrack to an office job, or modelling or book a flight back home to Rhodesia. It seemed ages before I was plucked from my reverie, hearing my name loud and clear. I rose from my chair and was ushered into a large carpeted room. The gentleman who took me into the room told me to wait at the back for instructions. I followed his advice and noticed the long conference desk facing me, behind which there was a sea of faces. These five interviewers sitting in front of me were going to determine my fate.

"Come forward," called one examiner, "Place your handbag and gloves on that small table over in the corner, then sit on the chair in front of us."

As I glided across the room, I felt all eyes following my every movement. Then, after placing my things on the table, I turned, walked towards the chair, and sat down as instructed.

After briefly introducing myself, then came the onslaught of questions: "Do you drink? Play any sport? Do you dance? How many hours does it take from Venice to London by train?"

Having told them during the introduction I had recently travelled by train across Europe to England, they asked how long it took.

"Have you ever studied psychology? How many languages do you speak? What minerals do the Northern Rhodesian mines produce? Who is the Chancellor of the Exchequer?"

They chose endless questions to suit the applicant. I was uncertain about the Chancellor of the Exchequer because I hadn't read a newspaper over the past week or two. So, I told the board I didn't know. They all smiled kindly, and one of them gave me the answer. "It's the Right Hon R. A. Butler."

"Thank you," I whispered.

Finally, an examiner handed me a piece of paper across the table. "Will you please read this briefing from the rear of the room."

I thanked him as he passed it to me, then got up and walked to the back of the room. Glancing at the paper, I realized it was a pre-flight briefing given by stewards and stewardesses.

"Imagine you are on an aircraft and there's a tremendous roar of the engines," one of the males said. "So that everyone can hear, you must speak loud and clear."

The room was so tense I could hear my heart thumping in my chest and my eardrums throbbing. I heard nothing else apart from the silence surrounding me. Scanning the paper in my hand, then holding it stable, and looking at the words, I recited:

"Ladies and gentlemen,

May I have your attention, please?

Our flight time to Rome will be one hour and 50 minutes.

Captain Jones will inform you on the progress of the flight from time to time.

We will fly at 40,000 feet, but the cabin will be pressurized down to an altitude of 8,000 feet so you will not feel any effect of the height.

The ladies and gents' rooms are in the rear of the plane..."

I read it to the end.

It was essential to place emphasis on certain words and phrases throughout. As I reached the end of the briefing, someone said, "Thank you, you may go. Leave the paper on the table in the corner. You will hear from us in due course."

I nervously placed the briefing paper on the table and picked up my handbag and gloves. Feeling a little light-headed and deflated, I thanked the panel and left.

One name I learned on the panel of interviewers was that of Diana Furness, the chief stewardess of the Comet Fleet. When I returned to the reception area to collect my jacket I had left folded over the back of a chair, there was complete silence. A few more girls had arrived and were sitting in suspense. They looked up at me, but no one smiled. They were probably feeling as empty as I felt at that moment. As I was leaving and about to walk down the corridor, the gentleman who was ushering us into the conference room summoned me to one side. He revealed I had got through the initial interview. However, the company would confirm it in writing.

"But how do you know?" I asked, stunned.

"That's my prerogative," he said, smiling.

I could hardly believe my luck. I felt as though I was floating on a cloud. But the elevator would take me to ground level and bring me back to reality.

~

MY SECOND INTERVIEW took place on Friday, the 1st of August, again at Airways House in Brentford. This time round there were four examiners who, once again, pounded me with one question after another. At the end, again they told me I would hear from them in due course.

The periods between interviews and letters from the airline seemed to be torturously long and drawn-out. So, in the interim I tried to gather as much general knowledge as possible as well as continuing my French classes. Anything that would help to get the job.

Days later another letter arrived informing me I had been successful in the second interview. Providing I passed the medical exam on the 14th of August, they would accept me for training. From 4,000 applicants, a hundred were chosen in the first round, and only eight successful girls attended the medical section of Airways House. We were all called together for rigorous physical examinations. Only if we passed these would we receive the many inoculations needed by the airline to enter distant countries.

BRITISH OVERSEAS AIRWAYS CORPORATION

Miss E.J. Baker,
94a, Cromwell Road,
London, S.W.7.

London Airport,
Feltham,
Middlesex.

5th August 1952.

Dear Madam,

offering you employment

1. I have pleasure in ~~confirming your engagement~~ on the staff of the Corporation on the following terms:-

Occupation **Stewardess.**   Grade   **Under Training.**

Salary   £ 5-0-0d   per week ~~xxxxxxxxxxxxxxxxxxxxxxxxxxxxxxx~~ ~~xxxxxxxxxxxxxxxxxxxxxxxxxxxxxxxxxxxxxxxxxxxxxxxxxxxxxx~~ ~~xxxxxxxxxxxxxxxxxxxxxxxxxxxxxxxxxxxxxxx~~. In addition while allocated to a flying roster you will receive £2.0.0d, per week flying pay.

2. Your date of engagement is **22nd August 52 at 9.30am** Your base on engagement is **Airways House, Brentford, Middx.** reporting to **Catering Manager.**

3. This engagement is subject to your references being satisfactory to the Corporation and can be terminated by one week's written notice on either side during the initial training period. Thereafter, one month's written notice must be given whilst based in the Continent of Europe, including the United Kingdom, or three calendar months' written notice whilst based outside the Continent of Europe.

4. Your employment will be governed by the Agreement dated 9th February 1949, reached by the Catering Panel of the National Joint Council for Civil Air Transport and by the Regulations of the Corporation at present in force and as amended from time to time. You will be deemed to be aware of all such Regulations whether or not you have in fact read them. Some information with regard to these is set out in this letter, details of others can be obtained from past issues of the Corporation's Gazette available for inspection at all Official Notice Boards, and from notices published by the Corporation on all Official Notice Boards.

5. If you have not already done so it will be necessary for you to satisfy the Corporation's Medical Officer as to your fitness to perform the above duties and if so you will receive instructions regarding an appointment.

6. It is a condition of this employment that you are accepted by the Guarantee Society. Enclosed with this letter is a proposal form which should be completed and returned to this office. The premium on this insurance is payable by the Corporation.

7. Subject to satisfactory service in the meantime, it is a condition of this engagement that you become a member of the Airways Corporations' Joint Pension Scheme on **1-9-53.**

8. Please indicate your acceptance of this appointment by signing the declaration on the reverse of the top copy of this letter and returning it to me. The second copy should be retained for your reference.

Yours faithfully,

L.A. POTTER, for Staff Supt.
(Recruitment).

Letter of employment Page 1

GENERAL CONDITIONS OF EMPLOYMENT

(a) Annual Leave   The leave year runs from 1st January to 31st December, and carry over from year to year is not permitted. For leave entitlement, see the Agreement dated 9th February 1949.

(b) Change of Address, etc.   It is essential that the Corporation is advised immediately of changes in address, marital status and next of kin, as considerable delays can be caused in an emergency if this information is not correctly recorded.

(c) Annual Salary Revisions   Increases in salary are not automatic, but are dependent upon satisfactory reports and are made effective on the 1st January, 1st April, 1st July, and 1st October, whichever is nearest to the date of engagement.

(d) Promotions   Promotion to a higher grade is dependent on ability and to there being a vacancy in such a grade. Such promotion and accelerated increases may alter incremental dates as if the promotion were in fact a fresh engagement.

(e) Insurance Cards   National Insurance cards should be handed to Heads of Branches or to Pay Offices during the first three days of employment.

(f) Inventions.   The Corporation, under its conditions of employment, reserves certain rights in any inventions or discoveries made or acquired by its staff (whether Patentable or not) which are capable of use in connection with the actual or potential activities of the Corporation or any of its associated companies.

It should therefore be noted particularly that all members of the staff should, before taking steps to patent any invention, refer the matter to their Departmental Head, who will advise them on the correct procedure to be followed.

SPECIAL CONDITIONS (IF ANY) APPLICABLE TO THIS APPOINTMENT

The rate of pay shown overleaf is applicable during the training period.

DECLARATION

Declaration to be signed in the presence of a witness.

I declare that I have read and understand the terms and conditions as outlined overleaf and I agree to accept employment with the Corporation on these terms and conditions.

Signature of Applicant..............................   Date...................

Signature of Witness..............................   Date...................

THIS COPY MUST BE COMPLETED AND RETURNED IMMEDIATELY.

Letter of employment Page 2

Looking around at the various faces, I was thrilled to see Jo among the group, and another girl, Myrna who was five-foot-six, slim and always stylish. She had dark, wavy shoulder-length hair, soft, kind eyes and an easy, cheerful smile.

Jo and I discovered that our flats were a walkable distance from each other. She lived in a flat with family friends, Mabel and Ossie Sykes, who owned a photography business that occupied the ground and first floor of the house, while they lived in the floors above. Jo had a top-floor bright and spacious flat.

"As we're now both working for BOAC, why don't you think about sharing my flat? You can have the front room overlooking Old Brompton Road," she suggested. "Mabel and Ossie are a wonderful couple and the greatest of landlords."

"Sounds like a good idea. I'll mention it to my flatmates when I get back home."

That evening I broke the news to my friends that I was considering moving to another flat to join my new co-worker since it will be more convenient with irregular working hours. The girls were sorry to see me go but understanding. That weekend, I gathered my belongings and hailed a taxi. As I closed the door behind me at 94A Cromwell Road, I felt a glimmer of sadness having shared many happy times with my friends. But Old Brompton Road was the beginning of a new chapter in my life.

BOAC TRAINING SCHOOL

O n the 22nd of August, we started the first day of our training. Jo and I set our alarms for 5:30am, as the sun rose, fearful of being late. Reeling with excitement, I tossed and turned all night, then I noticed the dawn lighting the sky. Crawling out of bed, I crept to the bathroom, and knocked on Jo's door as I passed, ensuring that she too was awake.

Still blurry-eyed and groggy, I looked at myself in the mirror and shocked at how tired I looked. Splashing my face with icy water, I lowered my head over the basin. It made me shiver and my cheeks tingle, giving them a pinkish glow. After a quick coffee and toast, Jo and I took an unhurried stroll to South Kensington tube to catch a train for Southall. We arrived at Southall in ample time for a connecting bus to drop us off at the catering section of the airline training school, housed in a cluster of austere redbrick buildings that used to be an old convent. No one knew why the place was abandoned, but BOAC took over the establishment as it proved to be a perfect location for crew training.

That morning, 14 of us, eight women and six men, gathered

together in an allotted classroom. Jo and I stuck together. The remaining six females dotted themselves around the classroom among the males. Everyone appeared to have selected their particular friend or friends, even at this early stage.

Wally, Mary, Ken, Myrna, John

When our instructor came in 15 minutes later, everyone immediately stood up. "Good morning, please sit down," he said, surveying the room. A short, stocky man with a ruddy complexion and small beady eyes, he looked smart in a gray suit with a pale-blue shirt and red-and-blue striped tie. Once he introduced himself as Mr Matthews, he asked our names and opened the lecture with the words, "Now, I must mention that punctuality is important. Do you all hear? I repeat, punctuality is very important."

There was silence throughout the room when one student muttered in reply, "Yes, sir."

Then he continued, "The training course lasts eight weeks. Friday of each week is a test day on the different subjects. You must reach an average of 85 percent to pass. Anything less, we may dismiss you."

We all sat at our desks, absorbing every word, our eyes fixed on our instructor. He opened a notebook and said, "We will cover the following subjects, please make a note of them. Aircraft familiarization, menu definitions, passenger handling, bar dispensation and catering. These will also include aviation medicine, air-sea rescue, jungle, desert and arctic survival."

This wasn't the end. Everyone documented the headings of the next 40 lectures in their notebooks. Mr Matthews chalked them on the board. "Stewards and stewardesses will have to deal with anything from a pin-prick to the birth of a baby, which we will teach you in the aviation medical course. Dr Phiffer, the company doctor, will train you on this subject," he said, then pointing to notes on the board, "The standard expected of stewards and stewardesses will be referred to throughout the course."

Mr Matthews stressed that our voice must be clear and carry to passengers seated in the rear of the cabin. We had to have powerful arms and hands as sometimes there is a certain amount of heavy-lifting to contend with. A flawless complexion, tidy hair and a tiptop appearance was always essential. The notes in our instructor's book were general for all intakes of stewards and stewardesses.

Besides pages and pages of written work, there were weekly examinations, along with periodic tests of endurance that included scrubbing, peeling potatoes, polishing and other arduous tasks. My friend Anne told me a steward had to peel a sack of potatoes while on her course. He refused and was dismissed. One must learn to carry out instructions on the spot without questioning the instructor.

With his hands linked behind his back, Mr Matthews paced the floor and continued with his lecture. "High standards of cleanliness and personal hygiene are essential. Clothing should be changed as often as possible in the tropics. To deal with many different passengers and events, a pleasant disposition is enhanced with common sense and tact. All these attributes are very important."

We all jotted his words into our notebooks when he continued, "You must have keen eyesight, astute hearing and a sense of smell. Honesty and sobriety are essential. As a matter of routine, each day begins with nail inspection and may I remind you that your nails must be short and free from varnish. Hair is to be short and off the collar. Jewelry is not allowed, apart from a wristwatch." There was a slight pause and then he said, "Any questions?"

My list showed the topics from inauguration to termination and I thought, 'One man in his time plays many parts', a line from Shakespeare's *As You Like It*.

He concluded the initial stage of the presentation. It took three hours or more intermingled with questions and answers. "You may now take a 45-minute break, but make sure you're back in the classroom on time."

As soon as our instructor left the room, we all rushed out to buy coffee and snacks from a comfortable tearoom across the street. Everyone had returned by the time our tutor arrived.

"Besides the academic side of training, there is also the practical that will take place in a mock-up. It is the grounded plane that you may have noticed in the convent's quadrangle," he said. "This mock-up is complete with cockpit, galley, passenger seats and toilets. It is in this confined space that you will practice handling trays. Also using cutlery, food and drinks, without the interactions with passengers. Oh, and you will learn about all five passenger-aircraft in the BOAC Speedbird Fleet. They are Argonaut, Stratocruiser, Constellation, Hermes and the Comet Jetliner. At the end of the training, you will be posted to any of these aircraft."

And so, he rambled on, though we found his words beneficial and absorbing.

~

THE FIRST WEEKEND CAME, I relaxed in bed and meditated with my eyes closed, thinking about the past week. Still in a state of drowsiness, I heard the shrill whine of jet engines overhead. This could be a dream. I swung out of bed heading straight for my window, and threw it open, letting the crisp morning air into the room. This year, summer and fall had been enjoyable with little rain and sporadic gray skies. But now the weather had begun to get colder and crisper, though we still had an odd day of sunshine, and this was one of those days.

It was a glorious morning, there wasn't a cloud in the azure sky. It gave a cheerful appearance to the grey chimney pots lining the London rooftops. Gazing out, I saw the most magnificent plane I'd ever seen, like a mighty swallow hovering motionlessly. The graceful lines of the fuselage and the swept-back wings glistened in the sun.

"Jo," I called out, "Come quickly and look at this stunning aircraft. It's the Comet. Oh! I would like to be posted to that fleet." It had disappeared by the time Jo reached my window. As I looked out at the empty sky, I clasped my hands together and prayed in silence: *"Provided I'm not dismissed from the course, please God, please, please, could they post me to the Comet Fleet?"* Then I realised God must be answering other peoples' more important prayers. So, perhaps my humble request was not heard.

Jo smiled and said, "Hope your prayers are answered."

"Which fleet do you want?"

"I don't mind really.... I have no preference," she said and returned to her room.

Little did I realise that an object such as an airplane could send waves of cold shivers through my body, giving me an inward sickness of pleasure. I made a point of enjoying every moment of the weekend since our training positions at the airport were allocated for the following weeks.

"Jo, would you like to take a boat ride down the Thames? We

may not have time to ourselves after this weekend."

"I'm going to Glastonbury to visit my mother and father," she answered, "Why don't you come along?"

So, we ended up spending the weekend in Glastonbury with Jo's parents, who were overjoyed to see her and welcomed me warmly.

We returned to Old Brompton Road on Sunday evening, relaxed but exhausted. On reaching the flat, Mabel and Ossie invited us into their lounge for tea and a snack. Afterwards, since we had an early rise, we were in bed by 10pm. I fell into a deep sleep for several hours, but during the night the ringing phone woke me. Glancing at my clock on the side table, it was 3am. I thought it might be bad news from home, but someone answered the phone because the buzzing stopped.

As I turned over to go back to sleep, there was a tap on my door. It was Ossie who was grinning from ear to ear, with eyes smiling behind his round glasses. "Elaine, Anthony Steel wants you on the phone."

A little dazed, I sat up reluctantly. "It's 3am," I said, rubbing the sleeplessness from my eyes. "What on earth does he want this early in the morning?" And I shivered; shivers like those when one has a malarial attack. I remember the symptoms well, because of having the sickness as a young child.

"Ossie, I can't speak to him, please can you take the call."

"I'm afraid, I can't. He has asked for you. You must speak to him."

I crawled out of bed. Ossie returned to the phone, and I heard him say, "Elaine will be down in a few minutes."

Feeling a little uneasy, I staggered down the stairs drowsily. I took the phone from Ossie, shaking as I held it in my hand.

"Calm down," he whispered to me, with a glint in his eye.

"Hello, Elaine," he said down the phone and, as always, his voice was mesmerising. He came straight to the point, "I'm having a party and would like you to come and join us. Errol Flynn is here. We're currently making a movie together."

"I'm sorry I can't. It's three in the morning."

"I know it is, but I live around the corner, close to your place, and we would enjoy your company. I'll send a taxi for you."

"What time do you go to bed at night?" I asked.

"I either go to bed late at night or get up very early in the morning," he answered.

"Well, I appreciate the invitation, but I have to be up in a couple of hours to head for airline training school."

We prattled on for a few more minutes, then he said, "So, you won't come?"

"I'm sorry, it's not possible."

"Well, some other time."

"Thank you anyway. Goodnight," I said and put down the receiver. Our conversation was brief but friendly.

By this time, I was trembling and could hardly walk up the stairs because of the cold and the thought of this movie heartthrob contacting me. As I passed Mabel and Ossie's room, they called out, "Are you all right?" and chuckled with me.

"Yes, thanks. I'm frozen and going back to bed."

~

ON ARRIVAL at the convent on Monday, while waiting for the first lecture, we, the girls, were standing in a group chatting about our weekend. After listening to their ventures, I told my 'three am' story. A very sexy, well-spoken colleague said, "So did you go?"

"No, BOAC means more to me than a party, even with a movie star."

"Oh! Wish we'd had the opportunity," they all said, sighing.

At that moment, our lecturer, Mr Matthews walked in, so we split up and rushed to our desks. After the usual morning greetings and nail inspection, he informed us we would be going to London to an exclusive beauty salon in Mayfair and would be there until mid-afternoon. Little squeals of delight burst from us all. Alternative duties were given to the males in our group. Then someone said, "Oh, please, sir, can you tell us which salon?"

"It's a surprise. We use the products, men's and women's, on all our aircraft."

The prospect of a diversion from our routine thrilled everyone. All preliminaries over, we boarded a small company bus that took us into the West End. Snaking through traffic-choked London, we made it to the salon, which was a hive of activity. The receptionist lost no time assigning us to our various cubicles. In the beauty treatment rooms of the salon, we were going to learn how to apply makeup that looks natural, have our hands and nails treated and our hair restyled.

They showed me into a booth equipped with subtle lighting, large mirrors and a reclining, comfortable bed. As soon as I relaxed on the bed, any tension in my body soon melted away. The lovely floral aroma and the softness of the masseuse's hands working on my face and neck were pure luxury. Another assistant massaged my nails and hands. Having two assistants working together helped in finalizing their tasks. After the facial, the beauty therapist had applied my makeup too. She held a mirror in front of me so I could see her artistry. It was glamorous enough but I didn't think it suited me. Growing up in a rural town in a hot, tropical country, I was used to a more natural, bare-faced look, so felt the makeup was too heavy and bright. It was perfect for someone on a movie set. I was then whisked away to the hairdressing salon. In front of a mirror, bordered with brilliant lights, I sat in an upright swivel chair. This gave me time to examine my made-up face.

The stylist appeared. She untied my chignon, my long hair sweeping down, held my head, gently turning it in various directions while looking at my face. The usual tactic with all hairdressers. "Are you sure you want me to cut your hair?"

"Yes, we have to wear our hair short according to the rules and regulations of the airline. It must be collar length."

She sectioned my hair with clips, and with an ultimate

gesture and a sweep of the scissors, started cutting it from underneath at the back. All hairdressers seem to resort to the same gestures before severing one's hair. She was meticulous and took well over 30 minutes to finish my cut. I was thrilled with the results, the short, sexy gamine crop that was fashionable, showed off my cheekbones and suited my face. Some famous movie stars, like Audrey Hepburn and Marilyn Monroe, were wearing short waves and pixie styles. Emerging from our cubicles, we waited for one another in the foyer, commenting on one another's made-up faces. After leaving the salon, we searched for a nearby ladies-room to wipe away our heavy makeup before returning to the training school.

During the following grueling weeks our weekends were not our own. We spent Saturdays and Sundays from noon until 11pm at the airport waiting on tables in the restaurant with experienced servers. We also helped with washing-up cutlery in the kitchen. This was to acclimatize us to work with, and talk to, different people and cultures. No matter what tasks we were given, nothing and no one was to get us down. We had to prove our ability to keep calm under all situations after long, grinding hours. From time to time, with the late arrival of an aircraft, one met very disgruntled people plus harried mothers with crying babies in their arms. Fortunately, the positive and friendly travelers outweighed the unhappy ones.

Before joining an airline one is inclined to believe that a hostess job is glamorous and romantic. Any such thoughts quickly disappear once arriving at the catering school.

Jo and I happily boarded our bus at the end of each day, then took the train to South Kensington. Once we were back at the flat, it was sheer luxury jumping into a hot bath in the evenings to relieve aching legs and muscles. I usually went after Jo, so I could immerse myself in the water and soak for almost an hour. Then, in our dressing gowns, we would cook together and carry our supper into the lounge on trays, while unwinding and chatting over the events of the day. After swallowing some eggnog and feeling quite satisfied, we did the washing up and fell into our beds.

The last two weeks of our course were the most exciting, when they sent us to be fitted for our uniforms, sculpted to fit and reflected the glamor of this golden-age of flying. While waiting to receive our wings was imminent, we were all concerned with ethics. We could still be let go if we were considered unacceptable. We lived in fear of being thrown out for any minor infringement during the eight-week training period.

One afternoon during mock-up training, they assigned me to serve aperitifs to passengers before the main meal. But first I had to stand at the front of the cabin and start delivering the departure briefing. We had to use the correct inflexions and sounds taught during the many occasions we gave briefings to our make-believe passengers.

"Ladies and gentlemen, (pause)

May I have your attention, please?

We are about to leave for Rome.

Our flying time will be one hour and fifty minutes.

Captain Jones will tell you of the progress of the flight from time to time.

We will fly at 40,000 feet, but the cabin is pressurized to an altitude of 8000 feet, so you will not feel any effect of height.

In the rear of the aircraft are the ladies' and gents' toilets.

There is a bar on-board, which will open soon after take-off, and we will be happy to serve you with any drinks and cigarettes you may desire.

As a routine regulation, I must inform you that there are emergency exits: 4 on the port side and 4 on the starboard side.

If there is anything you need, we will be happy to serve you.

Simply ring the bell, the button which is on the arm of your chair.

Will you now please fasten your seatbelts for take-off.

Thank you."

This was followed by a life-jacket briefing. We had practised these briefings over the past several tiring weeks and wondered if the life jacket would ever be enforced.

Then I went into the galley, collected my tray of empty glasses, and started down the cabin to serve the passengers with imaginary cocktails. Mr Lawrence, our chief instructor and Miss Scharffe, the catering instructress, sat in the middle of the cabin, taking notes on our service. A forceful nudge from behind, deliberate or unintended, was thrust into my back. My tray tilted and the few remaining glasses toppled into Mr Lawrence's lap. I was stunned.

"I am so sorry, sir," I said, looking at both Mr Lawrence and Miss Scharffe.

Our instructor handed the glasses to me to replace on the tray. "Try holding the tray a little more secure," he said, looking up at me. "Your feet might leave the cabin floor if the plane encounters an air pocket during a meal or drink service. It has happened and you need to balance the items in mid-air on the tray, particularly when your feet make contact again with the deck. Anyway, don't worry, you'll be fine with a little more practice."

Fortunately, I was not dismissed for this minor mistake. As I returned to the galley with my ill-fated tray, I glanced at Jo, who was also acting as a passenger. She kept her eyes to the fore and didn't even glance in my direction.

The final week of our training, during lectures, we were asked to speak to the class on any subject for three minutes. This meant quick thinking, as none of us knew when our names would come up. We were taking down notes on menu definitions when I heard my name called. I ran my hand through my hair as I paced between the rows of desks, wondering what I was going say. Since I came from a tropical country, sun, heat, life in the jungle, I wondered if one of these topics would be suitable? Then suddenly the light brightened, solar energy, a favorite essay subject at school. The cycle ran something like this: the sun's rays feed the earth, from there to the wheat, grain or grass, which is food for cattle, that man then consumes. And so, my points rattled on for the full three minutes. There was much to mention in the allotted time. Luckily, I noticed that my colleagues seemed to be listening and didn't show any signs of boredom.

To complete our training, we served a full-course meal of actual food in the mock-up to Mr Lawrence. He was always precise about 'dish and plate appeal' to make a meal appetizing. Also present was the catering manager Mr Belcher, Mr Matthews, our instructor, and Miss Scharffe, plus another two VIPs.

Dish and plate appeal

Training in the mock-up

Everything worked well and the enactment was a success. So, during this time, we had learned the correct way to carry out every aspect of our duties. In the model kitchen, they gave us meticulous instruction on the handling and preparation of food and presentation. By now we were wearing our tailored blue uniforms, which added to the glory of the practice.

Dorothy, Jean, Jo, Myrna, Daphne, Wynn, Elaine and Mary

On the last day, everyone gathered together in our classroom, awaiting final instructions. In pairs, we went into the chief instructor's office, where we received our staff number plus various other details. My number was 71079. Once all the formalities were over, a BOAC bus transported us to the London Airport office of the fleet catering officer, Mr Peter Drayson. He would post us to our various fleets. As I sat in the waiting area, I was full of apprehension listening to the names called. Each of us knew in which direction we wanted to go, although some may find the result disappointing. As our colleagues returned from the consulting room, we all looked up, and someone said, "Which fleet are you posted to?"

The different replies were, "Hermes, Argonauts, Constellations, Stratocruisers."

After 40 minutes, six stewards and eight stewardesses had been into the room designating their future. They assigned my colleague Jo to the Argonaut Fleet. Then my friend Myrna disappeared behind the closed door. She returned, grinning sheepishly as she elegantly crossed the room.

We all looked up expectantly.

"Comets," she replied.

"Oh how lucky," I said. "But how did you manage it?"

Finally, I walked into the room and sat opposite a stern-looking gentleman. He wasn't as stern as he appeared and was rather good-looking. After the usual formalities, he peered over his dark-rimmed glasses and asked, "Miss Baker on which fleet would you like to be posted?"

"Sir, that's up to you. You have my report and must know something of my capabilities, but given the choice I would like the Comet Fleet."

Reviewing the notes on his desk, he mumbled. "Hmm, Argonauts, Stratocruisers, Constellations, Hermes." Then there was silence. He glanced at me and said, "Comets."

"Oh, thank you, thank, thank you..." I was over the moon.

I'm sure he could see I was about to leap off my seat and give him a great big hug from sheer delight. But I regained my composure and shook his hand. My prayers had been answered. Before causing any embarrassment, he looked down at his notes and said, "Good luck and happy landings."

I made a hasty retreat to the waiting room where the others were standing ready to proceed to their different destinations.

"Well, to which fleet is our pocket Venus posted?" one of the male colleagues asked.

Looking at the ring of faces, breaking into a wide smile, I whispered, "Comets."

Everyone appeared happy with their designations. After sad farewells, we all finally split up on Monday, the 12th of October, to report to the managers of the various fleets. Myrna and I were told we had to present ourselves to the line manager of Comets for a week's conversion course. This would occur before we could be rostered on a supernumerary flight.

After the morning's event, instead of immediately returning home, we accompanied one another to Airways House to find out where we would meet for lectures the following day. Then headed for a nearby restaurant to enjoy a cup of coffee while excitedly chatting about our future destinations and all the places we longed to see.

COMET TRAINING

I woke up to bleak, bitingly cold morning on Tuesday, the 21st of October. My nose peeped over the quilt, and I looked at the gloomy gray skies outside my window. I shuddered, knowing that I would have to leap out of bed into the freezing room within minutes. Feeling enthusiastic about today's new venture, I dashed to the bathroom to get ready. The icy water splashing over my face to whisk away the sleepiness and puffy eyes. I wanted to feel bright and alert for our new intake of Comet 1 information.

Myrna and I arrived at Airways House about the same time. An eager attendant led us to the lecture room. There were six other people in the room, three stewards and three stewardesses attending a refresher course on the Comet. Our instructor, a tall, middle-aged man with salt-and-pepper hair and laugh wrinkles around his eyes, appeared in front of the room. His immaculate dark suit gave him an air of dignity.

"Good morning, everyone," he said, as he walked to his desk and introduced himself as Mr Sanson. "In the next few days, we have a great deal to cover. I'll start by giving you a little background on the De Havilland Comet 1, starting with excerpts from the recent BOAC Air News press review. If you wish, you can make brief notes and answer questions at the end."

The room was silent, everyone was interested and intrigued.

"The year 1952 will be a milestone year in aviation history as the Year of the Comet," he began giving us the history of the jetliner. "BOAC, as the world's first airline to operate a pure jet aircraft, gave Britain a lead over all other nations. A proving flight taken as far as Japan in the east and Johannesburg in the south. The Comet jetliner's fascinating tale started when de Havilland Aircraft Co. first explored the possibilities of using jet propulsion for civil flight. John Cunningham nicknamed 'Cat's Eyes' Cunningham, was a Royal Air Force night-fighter ace during the Second World War. He was also a de Havilland chief test pilot. On the 27th July 1949, he flew the world's first pure-jet-powered civilian airliner and rose to 10,000 feet. He flew the plane for 31 minutes before taking it down onto the runway to end a historic flight."

At the end of a short Q&A session, our tutor scribbled a few notes on the board and continued with the lecture. "Intensive test-flying took place immediately. And before the end of the year, the Comet made several overseas flights. In December, the manufacturers stated that the flying trials had shown that the aircraft could cruise up to 490 miles an hour. The normal operating height was 35 to 40,000 feet, nearly eight miles above the earth. Passengers are bound to ask about the wingspan, the length of the fuselage, the type of engines, so make sure you note all the details. As the aircraft is new to BOAC, those fortunate enough to fly on the Comet will hound you with questions related to the above..."

"Your next heading will be the Formation of BOAC Comet Unit." Then he continued.

"BOAC set up the nucleus of a Comet Unit in September 1950 to prepare for introducing the Comet on the Speedbird routes around the world. The personnel appointed to the head positions in the Unit were highly decorated pilots who flew with the RAF during World War 2."

There was a pause. Mr Sanson told us to take an hour break for lunch and be back in the lecture room by two o'clock. When everyone returned, he began the second-half of the lesson relating to overseas development flights. In the event of unusual questions being asked by the passengers, the cabin crew had to be knowledgeable about the Comet. If we didn't know the answer, we had to refer to the captain to gain the correct information for the benefit of the traveler. "You must never tell a passenger that you 'do not know'," said our instructor. "Say you will speak to the captain and return with the correct answer.

"The first overseas development flight carried out by the BOAC Comet Unit left London Airport for Rome and Cairo on the 24th of May 1951. This was piloted by Captain Rodley. The jetliner covered 927 miles between London and Rome in over two hours flying time. The Corporation made 12 overseas development flights. One of these was to Johannesburg and another to Singapore. The aircraft also visited many airports in the Middle East, the Persian Gulf, Pakistan, India and Africa. Before contract delivery, de Havilland loaned one of the production models to BOAC for their crew training. They started a series of training and familiarization flights on the Johannesburg route in January 1952. These flights gave ground staff at intermediate stops the opportunity to practice refueling. This and other procedures had to be carried out on the ground as quickly as possible. Other important facts were data collection for planning and confirmation of operating techniques.

"The pattern of a routine Comet flight is threefold," Mr Sanson continued. "First, the climb which takes 35 minutes to the most economical cruising altitude for jet engines, which is about 35,000 to 40,000 feet. Then, by fuel consumption, the cruise itself is a gradual increase in cruising capacity. Finally, the descent which begins some 200 miles from destination."

With a piece of chalk in his hand, he ambled back and forth in front of the blackboard. "Now, the importance of weather forecasts. This Comet unit has paid particular emphasis to the issue of meteorological conditions at high altitudes. And the prediction of these conditions by various authorities along the Comet routes. Flight-deck crew must take forecasts of wind direction and strength so they may plan the duration and other details of the flight with accuracy. The Comet unit has found the prevalence of jet streams an interesting but not a serious problem. These are streams of air traveling at top speeds, sometimes at over 150 miles an hour. Those encountered by the Comet have found the jet stream relatively narrow and soon traversed. They do not associate turbulent air conditions with them. In fact, the Comet offers a much smoother flight with less vibration than has been possible with piston-engine airliners. Aircrews must have accurate knowledge of the location and direction of jet streams to enable them to prepare their flight plans."

Our instructor returned to his desk, sat down and flicked through a pile of papers. The class was silent, wondering what was next on the agenda. "We are almost at the end of this session," he said. "It's important that you have noted the key details, so let's proceed. Your next heading will be, 'Opening of the World's First Jetliner Services'."

He then continued, occasionally reading from his documents. "There were 17 Comet familiarization flights between London and Johannesburg. This was before achieving the significant milestone in civil aviation. BOAC launched the world's first regular passenger jetliner service on that route on the 2nd of May 1952. The Corporation had by then taken delivery of four revolutionary aircraft. A fifth, which they previously loaned to the Corporation for crew training, was delivered later in the month."

Inaugural passenger flight 2nd May 1952 G-ALYP

Then our tutor resumed. "During May, there was one Comet service a week in each direction between London and Johannesburg. In June, they increased the frequency of this service to three a week in each direction. At first, the route was through Beirut in Lebanon. Then, as the unsettled state of affairs earlier in the year, improved in Cairo, normal operations of two of the services resumed through Egypt. This had the effect of still further reducing the scheduled times. Today, the overall time for the London–to–Johannesburg flight via Cairo is 21 hours and 40 minutes, only 17 are spent in the air. The Comet route to Johannesburg is through Rome, Cairo or Beirut, Khartoum, Entebbe and Livingstone."

Our instructor drew a map on the board, showing the route from the UK to Johannesburg and then continued the lecture. "On the 15th of May 1952, the Comet Fleet launched many training flights to Singapore via Pakistan and India. These incorporated Ceylon on the return journey. On the 11th of August, BOAC started a weekly Comet service through Rome, Beirut, Bahrain, Karachi and Bombay, along the 6,000-mile route to Ceylon. The flight to Colombo occupied a total of 20-

and-a-half hours, the flying time being 15-and-a-half hours. Ceylon, like South Africa, was brought to within well under a day's journey from London. BOAC inaugurated the world's third jetliner route on the 14th of October 1952. The reduction in flying times on routes by the Comet gives air travellers a new space-time concept."

After the lesson, they took us to the airport to view the interior of a Comet. I walked up the aircraft stairs into the attractive vestibule at the rear of the main cabin. Inside this beautiful aircraft, I was mesmerized and thrilled.

Luxury inside the Comet

The vestibule was flanked by a wardrobe for passenger coats and small baggage. Elegantly built in lightwood paneling with full-length sliding curtains in the blue, representing the airline. I noticed on my right, that separate toilets for ladies and gentlemen were located off the vestibule, with the ladies on the port side. On the starboard side the gents' toilet was fitted with a washbasin supplied with hot water from an immersion heater. Plus electric-razor sockets and a mirror. The ladies' was equipped with a dressing table, seat and large mirror. The soothing pale-grey and color scheme inside the cabin was sophisticated and luxurious. Each seat had an individual reading light, together with a call button. I walked through the cabin, absorbing everything from top to bottom. The BOAC Comets were each designed to accommodate 36 passengers and a crew of six. Behind the flight deck was the galley, plus a spacious freight compartment. Our instructor took us onto the flight deck and gave us a summary of the Comet cockpit with its many flight panels. This was a little too technical for me, though it was fascinating.

THE FOLLOWING MORNING, Myrna and I met at Speedbird House entrance. We walked together to the designated examination room, chatting all the way, but feeling apprehensive about the forthcoming exam. Mr Sanson appeared and handed around the test papers. I skimmed through the questions, then read and reread them slowly and started writing the answers. The paper was gratifyingly easy, leaving me time to read through my answers. The week after we received notification that everyone had passed with flying colors.

From Rhodesia to becoming a BOAC stewardess

FIRST FLIGHT

❧

W hen I received my first roster through the post, I was exhilarated. The first thing I noticed was the word Singapore. A whirl of images flashed through my mind, sultry, sun-dappled jungles and steamy neon-lit streets, palm-shaded French colonial houses. The rostering gods were on my side. I rushed up back up to the flat to tell Mabel. "I'm going to Singapore," I shouted out as I raced up the stairs, "I'm so thrilled!"

She was happy for me. "Well, at least you will relax in the sunshine. Send a little back to us when you get there."

On 26th of October, the day before my departure to the East, the weather was bitterly cold. After gathering my chattels together, I caught a taxi to London Airport as we had to check-in the afternoon before the following morning's flight. On arrival, at the briefing room, I read and signed the briefing book. Making sure I was up to date with the catering notes and any other significant documents. Then I made my way to the catering department where I found Rodey, my steward, and Janie, my senior stewardess, about to check their equipment.

An initial flight is made under supervision, so that one is familiar with cabin routine. Thankfully, I was under the guidance of two very competent and helpful people. The Comets carry only two cabin crew, a steward and a stewardess. On this flight, Janie was my tutor. She was a proficient, quick worker, long accustomed to the routine. Before this, she worked on Stratocruisers and Constellations. She told me we must use every minute because of the short flying time between places. Our steward too was experienced and had flown for some time on various aircraft.

"We have a super crew on this flight," Janie said. "The pilot is captain Wellwood, first officer Pitu, engineer officer Jones and radio officer Griggs." Then shaking my hand, she added, "I know who you are. I'm Janie Todd, and this is Ronald Rodford, but we call him Rodey. Later, you'll meet the flight-deck crew. They're currently in the briefing room and meteorological office studying an analysis of the weather for tomorrow. A reminder, at all times in front of passengers, always refer to the crew members as captain, Mr and Miss."

Once our tasks were completed for our imminent flight, we all boarded a bus for BOAC Dormy House in Sunningdale, where we would spend the night. This was a special guesthouse for crew leaving on early flights. The following Sunday morning, 27th of October, I was awakened at 4:30am with a hot cup of tea and a couple of biscuits, which I downed and then dashed into the shower. I hurried and dressed, making sure I looked smart and immaculate. Before leaving for the airport at 6:30am and meeting the crew at Dormy House entrance to board the bus, I grabbed a bite to eat.

Once we reached the Comet G-ALYR, the standby cabin crew had already situated everything in their places. Racks displayed magazines, newspapers, books and stationery. In the front of each seat was an air-sick bag and route information. We knew exactly where to find things without having to look. Rodey rechecked the galley equipment, while Janie and I inspected the bathrooms. I organized the cosmetics in the ladies', which included creams, lotions and perfumes. Then the men's amenities in the gents' toilet. Our special bag was well-equipped and remained with us for the entire flight. It was the only thing we take off the aircraft with us during the slip stops.

Janie and Rodey thanked the standbys for their assistance and bade them farewell. They took their places on the aircraft, ready to receive the passengers who had strolled across the tarmac. I sat in the stewardess' seat at the back of the cabin, watching the procedure carried out by my diligent workmates.

Once everyone was seated, the steward closed the back door, then went up front to notify the captain that all was well. My colleague returned to the rear to collect a handbasket containing barley-sugar sweets and small packets of cotton wool. The sweets were for sucking to help any discomfort in cabin pressurization variance. And the cotton wool to plug their ears during take-off or landing in the event of pain. These might assist in any discomfort with pressurization. She handed them round to the passengers while checking seatbelts were fastened. When she reached the tail end of the cabin, she placed the basket on the stewardess' table in front of me. Scanning the area, I noticed that the passengers appeared at ease, so, on the whole, must have been seasoned travelers. I was warned that as flying is a rarity for some, sometimes we may encounter a wary, anxious passenger who had to be consoled. I observed the proceedings sitting in my seat at the back. I glanced through the window occasionally. I could never take it easy again once being a fully-fledged stewardess.

Janie walked up to the front of the cabin, turned to face her audience and delivered the departure briefing. She returned telling me to give the life-jacket briefing. I wasn't expecting this, so initially felt a little anxious. When I reached the front of the cabin with a life jacket in my hands, my nerves seemed under control.

Finally, once the steps were moved away, the engines had been started, and G-ALYR departed and taxied along the runway. The airport buildings flashed by as we gathered speed and, with a last thrust, we were airborne at 9:28am. As we flew towards the channel, the panoramic view of the vivid-green English countryside fascinated me, unfolding beneath me. There were rivers, woods, beautiful farms and scenic fields. We were so high up, the houses and grazing sheep, cows and horses looked like little toys.

Janie grabbed my attention. "Make the most of this sector and absorb all you can. It's less than about two-and-half hours to Rome. After this, you take over with Rodey."

Then the seatbelt lights went off. Passengers began fidgeting in their pockets for their cigarettes. My colleague, unfastened her seatbelt and collected the complimentary cigarettes from the galley, handing them to the delighted passengers. She then made the round with newspapers and magazines, chatting to her flock and making sure they were comfortable. Most were settling down to their literature while others gazed out the windows.

Meanwhile, the steward was organising tea trays in the galley for the 36 passengers and cockpit crew of four. Rodey's individual trays were prepared with two sky plates. One with delicious-looking cakes, the other with several sandwiches and a steaming cup of tea.

Janie approached me. "Miss Baker, there's another tray of sandwiches in the galley to be distributed. Please would you hand them round to the passengers?"

The display was appealing, and one passenger remarked, "These look fantastic. Do you mind if I take three or four?"

"Of course, you're welcome."

As I reached the rear of the cabin, a red light flickered on the steward's lighting panel. My workmate, still busy with the passengers, asked me to go up to the flight deck and find out what the commander wanted. I approached the captain, who was looking at a chart handed to him by the first officer, then sensing my presence, instinctively turned around. "Miss Baker, please tell the passengers we can see Mont Blanc in the foreground on the starboard side of the aircraft."

I whirled back to the cabin and started relating the information to everyone. The passengers stared in wonder through their portholes at the panoramic view of the snow-covered Alps. Now, at the back, I turned to Janie and said, "Well, traveling at this speed Mont Blanc must be miles away by now and any mountain six miles below could be Mont Blanc."

The majestic ruggedness of the expansive range upon range of mountains was covered with a mantel of bluish-white snow.

After the tea service was over, my colleague gave me a report to complete. It not only stated the number of passengers but age, nationality, religion and personal information. It was important because those with special ethnic or vegetarian diets had to be catered for. Before leaving London airport, diets were normally organized if we were carrying an ethnic group. We passed the information to takeover cabin crew along the route. Everyone had to be considered and satisfied, including babies too young to have solid foods. Here, we uplifted special baby foods. I always had to mention on the report to anything unusual that might have occurred during the flight.

As the aircraft circled the runway on approaching Ciampino Airport in Rome, the steward's light panel flashed from the flight deck. So I went up front to find out if they required anything.

"As this is your first flight, we will let you watch the Rome landing from the cockpit," said Carl, the first officer, who had changed places with the captain and was landing the aircraft.

"Oh, how exciting," I said, viewing our approach thousands of feet above the ground.

As the aircraft descended, captain Wellwood pointed, "Look over there at the Comet that crashed yesterday on take-off. It's a complete write-off. Captain Foote, the pilot, has been suspended and will probably be transferred to York freighters. Fortunately, there were no casualties."

I looked at the wrecked aircraft at the end of the runway. "What happened?"

"We won't know until the board of enquiry issues its verdict," Carl responded. "But, my dear, we are currently flying a timebomb."

A strange comment, I thought, and it passed over my head.

Then, as the undercarriage doors opened, the wheels were lowered. We were gliding over a racecourse towards the runway at Ciampino Airport. When the Comet touched down and raced along the landing strip, I thanked captain Wellwood for the experience and returned to the cabin to help my co-workers with final chores.

Once the aircraft came to a standstill, Janie delivered the transit briefing. After unstrapping their seatbelts, the passengers gathered their belongings, and left the plane down the steps that were rolled up for disembarking and embarking passengers. I descended the stairs immediately after them and walked over with them to the airport building.

Some disembarking visitors were either remaining in Rome or had onward flight reservations. Those remaining were busy with arrival formalities while those in transit, snacked in the airport pizzeria. Janie and Rodey were left preparing the cabin for the next leg of the journey to Bahrain via Beirut. When I got back to the aircraft, Janie handed me the passenger list of the new arrivals boarding. "We have a VIP on this sector," she said. "It's world-renowned violinist Alfredo Campoli. He's travelling east to give a concert recital."

DIVERSITY EN ROUTE

$\mathcal{L}$eaving Rome, every seat was occupied. We arrived in Beirut at 6:05pm after three-and-a-half-hours in the air. On the flight, we served afternoon tea followed by canapés, cocktails and other drinks. After landing, I delivered the customary briefing. The passengers continuing the journey would have dinner at one of the Beirut Airport restaurants. On arrival at the eatery, I ensured that they were all accommodated at their chosen tables. Finally, joining my crew at their table, my captain told me to inform the passengers that there would be at least a 24-hour delay in Beirut. We had acquired a technical hitch which would be fixed the next day, so we would be spending the night at the Baalbek Hotel.

As I walked from one table to another delivering the news, they all seemed pleased about an unexpected stopover in an exotic locale. Then after finishing dinner, Janie and I boarded a large bus with the passengers. Rodey joined the crew transport after finalising formalities at the airport. This included removing the baggage that the passengers required for their unexpected stay. Then he called the hotel to confirm accommodation for everyone.

We drove through the bright lights and bustle of Beirut. In

the 1950s, the sudden influx of foreign money, turned the port city into a glamorous hub for affluent travelers with luxury hotels, pavement cafes and clubs springing up along the seafront. Then onto an open road, which started a slow climb, winding our way up twisting hairpin bends along roads lined with cedar forests.

In the dark, my eyes seemed to deceive me. I wasn't sure whether the trees were poplars, cedars or pines. After about an hour's drive, the bus arrived at a luxurious hotel in the mountains of Lebanon. Arriving at the hotel, telegrams were sent to friends and relatives about the delay. By this time the passengers were a little worn, but happily settled in the lounge and bar, enjoying cocktails. Janie and I walked around informing them of their room numbers and handing out the keys. It was well after ten in the evening before anyone retired.

The following morning heralded a gorgeous day of blue skies and sunshine, though there was a nip in the air. My room overlooked the main terrace of the hotel. Many of our early passengers had seated themselves on the sunny terrace before breakfast. A view from my other window presented a picturesque scene of rugged gray-green mountains. A tour of Baalbek had been arranged for everyone after breakfast. As Janie and I were concerned about the welfare of the passengers, we accompanied them on their expedition. After a short drive, we arrived at the majestic ruins in the early part of the morning, the crowning glory of Baalbek. The stone columns, pediments and massive walls depicting this Acropolis of time.

"This place was known to the ancient Greeks and Romans as 'Heliopolis', meaning City of the Sun," said Janie, acting as an impromptu tour guide.

"It's hard to believe that some structures are still standing after all the earthquakes and centuries of destruction," I replied.

We strolled and spoke with various people, going from one relic to the next, over piles of massive, sculptured stonework and broken columns.

As I wandered near the Temple of Bacchus, Mr Alfredo Campoli came up to me. "Miss Baker, please may I take a cine of you in front of that column. Your charming red-and-white floral dress will add contrast to the stone pillar."

"It's a pleasure, Mr Campoli, hope I do the picture justice."

He panned his cine camera from one side of the ruins to the other with me in the middle.

"Have you photographed the other important temples?" I asked. "Such as the Chief Temple, which may have been started in Nero's reign, sometime around 54 AD. Then there is the Temple of Venus near the acropolis that dates back from around 245 AD."

"You seem well-informed about the ruins."

"Well, whenever I stopover in a place, I always like to know a little about its history."

"This has been a pleasant interlude before my days of tight schedules. I'm thoroughly enjoying myself."

Temple of Venus, Baalbek

THE LITTLE CIRCULAR Temple of Venus was some 300 yards away so we walked to it through the whispering poplars. All the passengers were having an exciting day in this historical environment. As we picked our way through the ruins back to the main section of the Acropolis, I realized I wanted to unwind, so I took the hands of two male companions. The others also joined up, and we leaped from one giant limestone slab to the next, lying in the quarry.

Main section of the Acropolis

It was strange to see middle-aged men doing the things I still loved to do, their inhibitions thrown to the wind. For a few hours, we were free as birds. As we laughed and joked, I thought that tomorrow these men would return to their more serious side of life. The older members of the female sect couldn't compete with their male counterparts. They walked through the ruins in a dignified manner, but also seem to enjoy the release from the daily grind, laughing and chatting to one another.

After whiling away a few calming hours in this garden of antiquity, we boarded the bus for our return to the hotel. The eerie muffled call to prayer from a nearby mosque could be heard penetrating the evening air.

A section of the Garden of Antiquity

The next day, at 3:30pm local time, we left Beirut a little wiser for the unscheduled delay. Once airborne and everyone was settled in the coziness of the cabin, light refreshments of coffee, tea and biscuits, were served to the full group of contented passengers. It took about two-and-half-hours to fly between Beirut and Bahrain. Standing in the rear of the Comet, surveying this complacent scene, Janie said, "This is one flight we will always remember. I'll be sorry to hand this cheerful group to the takeover cabin crew.

"We've all had so much fun together and the passengers seemed to enjoy mingling with us," I replied.

As passengers were drinking two or three cups of coffee, with occasional snoozing, it was time to fasten seatbelts. Then tidy the cabin, gather newspapers and magazines, and organize things for the next crew.

"Miss Baker, will you please relay the transit briefing to the passengers," asked Janie.

"Certainly."

We didn't realize we were actually on the ground as the wheels touched down on the Bahrain runway. So, as we taxied to the bay, I walked to the front of the cabin, facing the passengers, began the landing briefing.

Jani and I helped the few passengers to collect their items from the overhead rack before they left the aircraft. But before any of them moved forward, we made our way to the exit to see them safely off the plane. They knew we were changing crew here, and as they disembarked, Mr Campoli passed by and shook my hand. "Thank you for an interesting excursion. I loved the interlude in Beirut." He handed me his business card and asked me to contact him in London if we were ever there at the same time.

"Thank you, Mr Campoli, it will be a pleasure. All the best for your eastern tour."

A ground-staff member met the travellers and escorted them to the airport building. Meanwhile, my tutors were handing over their reports to the ongoing steward and stewardess. "There's nothing unusual to relate, the passengers are a delightful group," said Janie to her colleague.

"Well enjoy your rest period here," she answered. "You may be invited to a sheik's party. They seem to take delight in entertaining aircrew."

"It will be a pleasure to have sunshine and swimming after the miserable weather in England," I said.

Refueling activity was taking place around the Comet. A fuel tanker rolled up, then a giant tube from a mobile air-conditioning plant arrived to pump cool air into the aircraft. Another truck appeared to collect baggage and freight. The BOAC ground-operations manager Stan Edge met captain Wellwood and his crew as they left the flight deck to discuss any current eventualities.

Formalities complete, we boarded the crew bus, where I sunk back into my seat and slipped off my shoes. We travelled about four miles in this old conveyance over a rough road to the BOAC guesthouse on Manama. On the way, we passed a kaleidoscope of scenes, including scruffy, skinny dogs that looked as though they hadn't had a meal in months. My heart ached seeing them. The odd donkey, without a tail or ears or both, pulling a cart laden with all kinds of commodities. The poor camels herded along looked so dejected. These unloved desert workhorses filled me with pity and sadness. They lived in oblivion, trudging along from day to day with nothing to lighten their burden. In fact, my heart cried for all the animals, the stray dogs, horses, donkeys and camels, were in a state of neglect and had such harsh, desperate lives.

At our simple and homey rest-house, after depositing our baggage in our rooms, we beelined for breakfast. Afterward, in preference to having a nap, we all changed into casual summer dresses and shorts with bathing suits underneath, and headed straight to a nearby beach.

The beach was very inviting with powdery-white sands, clear blue water and foamy white waves rolling onto the shore. Throwing our towels onto the sand and stripping down to our swimsuits, everyone dashed into the surf. Carl swam next to me, then when we were about 50 yards into the ocean, he began frolicking like a porpoise, ducking and diving. His head popped up, and with a broad smile, he said, "It's hard to believe we're being paid to swim in the Persian Gulf."

"I feel like I'm living a dream. The world is handed to us on a plate," I said, glancing around at the vast ocean. "But at this moment I'm a little wary about what might lurk beneath me, so I'm going to return to shallower waters." I struck out and raced to the shore, Carl tailing behind at a slower pace.

Back on the beach, Mr Jones, the engineer, produced a rubber rugby ball which he inflated. We spent the rest of the afternoon playing ball games and acquiring the sought-after tan in the scorching sun. Feeling parched from the exercise, we lounged in the shade of a little beach hut, over cold drinks, until our transport arrived to drive us back to the rest-house.

Experiencing the effects of the day's activities and the intense heat, I fell onto my bed after lunch and slept for a few

hours. A knock on the door woke me in the early evening. It was
Jo Stanbury, my flatmate, who was scheduled on an Argonaut
flight to the Middle East, and was 'slipping' in Bahrain for two
days.

"Hello Jo, how nice to see you. Have you had a pleasant
flight?"

"Not bad. We had an engine problem, which was soon solved.
However, the crew is going to a party this evening. Would you
like to join us?"

"Thank you for the invitation, but I'm going to take a shower
and have an early night," I said, "Let's meet in the morning and
walk along the waterfront."

"Okay, sounds good. I'll see you tomorrow.

THE FOLLOWING morning Jo met me in the foyer of the rest-
house. We made our way to the waterfront and strolled along,
taking in the extraordinary sights. Tiny shipyards were in
evidence where we watched artisans building dhows, crafted
from teak imported from Burma.

Elaine in front of a dhow

A suave turbaned Arab in charge approached us. "I notice you're interested in the building of the dhows." Then pointing to a hull that shone in the sun's beams, he said, "The hull is coated with fish oil that protects the wood from the seas and the weather. And the dhows ply the ocean to carry merchandise to the east coast of Africa, we also use them for pearl fishing."

Jo in front of a dhow

Dhow building yard

"It's fascinating to watch the artisans do this remarkable job," Jo said, smiling at the overseer who looked more like a ship's mettlesome captain.

"I'd love a voyage in a dhow," I said. "It would be something different and exciting." Instantly, my imagination ran wild. "And perhaps a pirate ship might attack us on the high seas. How romantic!"

We all laughed. The supervisor told us that dhows had been used in the Persian Gulf, Indian Ocean and Arabian Sea for over 2,000 years. They once carried ivory from East African ports to Kuwait and Abu Dhabi.

Turning to our new friend and advisor, I said, "Thanks for this enjoyable brief encounter, but we must continue our walking tour. We still have a lot of ground to cover."

"It's been nice meeting you." He shook our hands and said, "Come back soon."

"*Ma'al-salama*. Is my little knowledge of Arabic correct?" I called out as we began walking away.

"It will pass," he said, giving a parting wave.

As we continued along the narrow, gravel road, an Arab woman in purdah, black-robed with a veil covering her face passed by. Only her exquisite green eyes were revealed. On flat-heeled mules, she disappeared down a dark narrow alleyway. We stared after her, entranced.

"How striking that robe is!" I said to Jo. "I wonder if we could buy anything like that in one of the markets."

"Hmm... I agree. It would be perfect for one of our Arabic-inspired dinner parties."

Exploring the area further, we momentarily considered going into town, but Jo was reluctant. "The last time I was here I ventured into town, but won't do it again. Driving is dangerous and horns are continuously blown as soon as anything comes into sight. There is, of course, always something in sight."

"I've heard there are no traffic regulations at all, so I don't think its worth driving on the roads," I said. "Anyway, I'm impressed with Bahrain in the short time I've spent here."

When we got back to the guesthouse, Janie sprung up out of the blue and said to me, "Better get changed, the crew have been invited to a sheik's desert party tonight. It's an insult to refuse."

"Oh dear! Jo, thanks for an interesting day. It's been great fun. I'll see you back at the flat in Old Brompton Road."

Elaine and Jo outside the Rest-house

We parted company. In the airline, as the old saying goes, "we're all like ships that pass in the night, here today, gone tomorrow." Captain Welwood hired a jeep to cross the desert sands where the sheik was holding his sunset party. He acted as the driver and the whole bunch of us crammed into the remaining seats. It was a rough ride: up, down, sideways and over the dunes. When we reached the top of the final dune, the sun was casting a beautiful pink-golden glow across the sky.

Desert evening on the dunes

So, we stopped the vehicle on the crest of a dune to take several pictures. Then, gazing down into the shallow valley below, we saw a vivid pop of color glistening like an oasis in the middle of a desert. It appeared to be a large Bedouin camp. Nestled among the dunes, a massive tent had been erected for the evening's festivities, or possibly it was a permanent fixture for these deserts soirees. Many smaller tents encompassed the central conformation. Camels wandered around the periphery, making the site seem more authentic.

A Bedouin camp

"Goodness, it looks like a spectacular movie set," I said to my colleagues.

"Well, I'm no Peter O'Toole," said our handsome radio officer while laughing.

The Arabs were famed for their hospitality and a chance meeting often led to being welcomed with a glass of mint tea or invitations to parties in the desert. We all scampered back into the 4x4 and drove down to the fascinating location. Inside the tent, bowls of exotic fruit and nuts, figs, dates, papaya, pomegranate and persimmon, adorned the low-set tables. Persian carpets used as tablecloths and illuminated by candlelight added a magical atmosphere. Soft cushions for seating were arranged around the low tables. Several yards from the primary area stood a backdrop of small cube-like tents, each staffed by an Arab vendor. In one of the tents, was a well-behaved falcon perched on a pole with a tiny gold chain attached to its leg. He cast his glittering eyes at the scene in front of him and seemed excited when his owner took him into the middle of the tent to perform his swooping and soaring antics.

The desert at night, with the flickering candles and a velvety-blue sky full of stars, was hauntingly beautiful, while everyone succumbed to the niceties of a barbecued dinner. I had seen nothing quite like this Arabian Nights-style feast that was out of this world. Janie and I were given privileged seats on either side of a very handsome, middle-aged sheik adorned in his *keffiyeh* and flowing white kaftan. A shimmering dagger in an elaborately designed silver sheath was placed in his waist belt for effect. The rest of our crew were seated close by.

The first course was roast sheep, in fact, it was the whole sheep. We were all briefed that according to local custom, if you were offered something to eat, it was an insult to refuse. We knew also that sheep's eye was a local delicacy, so I avoided that by picking the fruit on the table.

Janie and the sheik were in a deep conversation when the sheik's attendant handed her a white cotton *keffiyeh*, which she put on her head. She picked up a corner of the cloth, draped it across her face, only revealing her large dark eyes, and said to him, "Well, how do I look?"

"Janie, you look more seductive than ever," he replied, amused.

I turned away and watched the belly dancers running onto the stage. The curvy dark-haired dancers, shimmying and shaking in wisps of silk and jangly gold jewelry, performed a riveting, hip-swiveling dance on a round platform in the center of the cosmic tent.

As the dinner progressed, a sheep's eye on a plate was passed around the circle of diners, but everyone avoided the

Janie's white cotton keffiyeh

presentation. When it finally reached our host, he looked at Janie, winked, and popped the eye into his mouth. You could, in fact, hear a sigh of relief pass through the air. All round, it had been a memorable and entertaining evening, and everyone was in a jovial mood. Janie kept the *keffiyeh*.

CREW CHANGE IN CALCUTTA

The endless shimmering of gold sands against the sapphire blue skies was mesmerising. Glancing through the porthole window as the aircraft left the ground, I thought this was a region of magic carpets and Arabian nights. I could live in this fantasyland forever. We carried a full load of passengers on the following three sectors to Calcutta via Bahrain, Karachi and Delhi. It was all work without five minutes to breathe. Before flying on to Singapore, we spent a couple of rest days in Calcutta. I was excited because I knew that there were places to explore. I hankered to see the Black Hole of Calcutta, which I learned about in a history class at school. And now I could view it in person.

In 1756, the Black Hole was a small dungeon in the old Fort William, where soldiers and civilians were held overnight in cramped conditions. Many died from heat-exhaustion and suffocation. When the door was opened the next morning, out of 146 incarcerated prisoners, only 23 survived. This was a hypothetical figure, but we will never know the exact number of survivors. The event took place a lifetime ago. Troops of the Nawab of Bengal held British prisoners of war in this situation after the capture of the Fort.

The joys of flying. Far-flung reaches of the world, intriguing, mysterious places that one had read and dreamed about in books, had become a reality. At the end of a smooth flight, captain Wellwood landed G-ALYU in the early evening on the 1st of November 1952 at Dum Dum Airport in Calcutta. We had been in the air for over seven hours from Bahrain. Then spent an hour on the ground, refueling at Karachi and Delhi. At both these places some passengers got off while new passengers joined our flight. At Dum Dum, as we handed the flight over to the next crew, the ongoing passengers were sorry to see us leave. It was late when we passed through the various formalities and boarded the crew bus outside the airport buildings. Once on the bus, I relaxed in a seat, took off my cap and kicked off my shoes, rubbing my tired feet. Janie, sitting on the seat opposite, did likewise, glancing at me and grimacing. We were all ready for the nine-mile drive from the airport to the Great Eastern Hotel in the heart of Calcutta City.

Sitting next to an open window, I breathed in the sultry night air and looked up at the bright moon surrounded by a blaze of shining stars in a clear sky. Oh, the beauty and serenity of tropical nights. The old bus rattled along at a slow pace, so I rested my head on the back of my seat, but every bump stopped me from drifting off. I longed for a comfortable bed.

Eventually, reaching the city, we passed through a poor neighborhood with narrow, curving streets, almost clogged at points with people and traffic. Even at this late hour, the beggars who had not already taken up their sleeping positions on the side of the roads were wandering around or sitting on street corners. On one of the sidewalks, in the somber lighting, I noticed several half-starved, stick-thin men, wearing tattered clothes, barefoot, squatting on their haunches. At a closer glance, they had bushy, matted, shoulder-length black hair. They stretched out their bony hands, begging passers-by to give them food or money. Others huddled together rolling and throwing stones into the air.

"Whatever are they doing?" I turned and asked the first officer, Carl, sitting behind me.

"Gambling with stones to see who will get the first meal of the day," he replied.

I had a lump in my throat, speechless. It made me deeply sad. The general squalor, filth and abject poverty were something that had to be seen to be believed and that you never get used to.

We pulled up outside the Great Eastern Hotel in Old Court House Street. The crew were already signing the register at the reception desk in the foyer by the time Janie and I put on our shoes, and collected our things. Stepping from the bus in front of the hotel, macabre figures covered from head to toe with white pieces of cloth lined the pavement. I stood transfixed, then noticed sections of the walkway scattered with red blotches that I thought was blood.

"Janie, is this normal?" I asked as we carefully made our way through the torsos so as not to disturb any of the sleeping figures. "Are all the red smatterings on the pavement blood?"

"No, that's betel nuts. These people masticate and spit anywhere and everywhere."

As I stood at the hotel entrance and slung the strap of my handbag over my shoulder, I said, "Oh! And look at those poor people huddled against their sacred cows lying in the road."

The animals held their heads high and looked around with lovely soulful eyes, not at all disturbed by passing traffic. They seemed contented to have found an allotted place to rest. My eyes welled up with tears. I was so saddened, so crushed at the picture before me.

Janie, seeing my misty eyes, touched my arm. "Don't worry, you'll get used to these scenes in time."

Once we had dropped off our belonging in our rooms, our crew and the crew returning to London met in the hotel lounge before going into dinner. I seated myself on a chair between Janie and Rodey, with captain Wellwood perched opposite. He looked across at me and said, "Elaine, you will not be continuing to Singapore with us. You will join captain Cudderford's crew and return to London tomorrow. His stewardess has fallen ill and there's no time to have someone flown out to replace her."

"But I haven't completed my training under supervision," I said, surprised and little disappointed.

"That's not a problem. One of the best and most experienced stewards on Comets will work with you, so there shouldn't be any mishaps."

I felt a little anxious about the situation. But both crews soon put me at ease. Bobby, the steward, walked up to me and patted my shoulder. "Don't worry, darlin', you'll be fine."

Then captain Cudderford intervened, "Let me introduce you to my crew as we will be working together for the next week. This is first officer Smurthwaite, radio officer Nixon, engineer Forster, and your partner, whom you've met, steward Bobby Smith.

"Hello everyone, I'm delighted to meet you," I said with a wide smile. Turning to captain Wellwood and my crew, I added, "Captain Wellwood, Janie and Rodey, thank you so much for your help and patience. I hope I do justice to everything relating to my position. Now if you will excuse me, I must retire and prepare for tomorrow's adventure. Goodnight everyone."

~

IT TOOK SIX DAYS, which included three rest days, for our crew to arrive back in London. On the final Rome/London sector, the flying time was estimated at three hours, owing to headwinds. We departed at 5:37pm from Rome and were scheduled to arrive in London three hours later. Bobby and I were given plenty of time to dispense canapés, cocktails and drinks before serving dinner to a full load of cheerful passengers. They appreciated everything presented to them. After the meal, when removing the trays, the passengers extended thanks for a wonderful dinner.

I passed around cigarettes as they sipped their coffee and liqueurs, and enjoyed playing cards and board games.

Before landing I was busy arranging coats in the cupboard, when Bobby walked over to me, smiling. "Darlin', your name will be in lights when we get back to London. You've worked well. I'm very pleased with you."

Feeling elated, I raised my eyebrows and said, "Bobby, it's been an absolute pleasure working with you." We called each other by our first names as we were out of the passengers' hearing range. I was glad that Bobby was happy with my work since it was customary for the steward to write a report on his stewardess. The captain was also required to draft a report on his entire crew.

The weather was dark and frosty by the time we landed on 8th of November. As I opened the aircraft door, a stinging blast of icy wind hit my face. Somewhat disheartened being back in this dreary climate after the glorious hot weather, I was already missing the sunshine and the swimming pools. Sometimes, with the constant flickering change of climates, I felt I was living another life in all these other places.

Once I had seen the passengers across the tarmac and into the airport building, I returned to the Comet to collect my cosmetics bag. This had to be handed back to the amenities service area. Bobby was busy arranging transport to the catering section for his bar. While the flight-deck crew had sauntered to operations for debriefing.

After going through customs formalities, I stood at a bus stop outside the airport building. Unexpectedly, an Allard sports car stopped, and a voice called out, "Want a lift to South Kensington? I'm spending the evening in the area with a friend."

It was Mr Forster, the engineer from our crew.

"Thanks a million. It'll save me taking the train from Hounslow. I live in Old Brompton Road."

"That's easy. My friend's place is in Queen's Gate."

When we pulled up outside of my home, Mr Forster assisted

me out of the car, took my suitcase from the boot and offered to carry it upstairs.

"Don't worry, I can manage. The ride is appreciated, thanks so much," I said, before waving him goodbye.

I unlocked the door of the flat, tip-toeing and towing my suitcase up the stairs I tried not to disturb Ossie and Mabel. It was fairly late, sometime around 10pm, their lights were off and they must've already gone to bed.

"Hello, is it you, Elaine or Jo?" Mabel called out.

"It's me," I replied, passing their door. "I've returned early. I never got to Singapore because I had to replace a stewardess who fell ill in Calcutta. The flight has been a wonderful experience."

"Well, it's nice to have you home. Make yourself a hot cup of tea and go to bed. We'll talk tomorrow."

"Thanks. Goodnight."

I trudged the three flights of stairs up into my vast, bright flat, and dumped my suitcase in the corner of my room, flung off my shoes and uniform, slipping on my dressing gown. I popped into the bathroom to take a long, hot bath to soothe my aching feet.

It was sheer bliss when I finally sunk into bed.

HISTORIC SITES

J o had not yet returned from her flight to the Middle
East. So, until she arrived back, I did my usual chores,
tidied the flat, washed my clothes at the launderette and
dropped my uniform suit at the cleaners down the road.
I always like to have everything organized within a day or two of
returning in the event of an unexpected assignment for standby
duties or for a flight.

As I had two full weeks off before my next trip, once Jo got
back, we wanted to go off on a little adventure, somewhere close,
so flew over to Ireland for a couple of days and visited Blarney
Castle, a few miles from Cork. Somewhere in history, we had
both descended from Irish ancestors. Apart from our
genealogical past, Ireland being across the water, not too far
from England, sounded appealing for a weekend escape. Instant
reservations were made taking the first available flight across the
Irish Sea.

When we arrived at Dublin Airport, we hired a car and took
turns driving the three hours to Blarney Castle. Passing through
sweeping green countryside with surrounding fields stretching
for miles, it was intriguing, gazing at the peaceful beauty of the

grassy green hills and dales on either side. Sheep were grazing in some pastures, and distant farms emerged dotted on the landscape.

"Jo, I can understand why Ireland is referred to as the Emerald Isle."

"Yes," she said with a smile. "It has a shade of green all of its own, the perpetual rain helps the grass keep its vivid color."

When we reached the castle mid-afternoon, we were lucky because there were a few tourists on this day. After strolling around the castle, an ivy-draped 16th-century tower set on impressive grounds with its lovely, lacey fern gardens, Jo and I climbed up the claustrophobic spiralling stairs to the battlements to kiss the Blarney Stone. This was partly our reason for flying across the Irish Sea. It is a known fact that kissing the Stone of Eloquence furnishes one with the 'gift of the gab', a legacy that will remain for seven years. Hmm.

"Jo, it won't be easy. But I should think having the gift of the gab for seven years is worth leaning upside-down backwards, over the parapet, and kissing the stone." I could see the concern on her face and felt the fear in my heart.

"I thought it would be a bonus until death do us part. Look at the distance from here to the ground, it's 85 feet! It'll be a disaster if we fall."

"Try not to think about it," I said cheekily.

After carrying out our mission, which was not too difficult, we walked down to ground level. Jo loved scrutinizing the castle and running her palms over the rugged old walls. These had withstood the elements across the milestones of time for centuries. Mission accomplished, we wandered back to the car and once in our ride, drove around looking for somewhere to spend the night.

Outside Cork, an idyllic country inn appeared on the scene. The B&B was run by a jovial, bald, plump landlord with a large, red, bulbous nose. He seemed to take an instant fancy to us and invited us to join him for dinner. We thanked him for his hospitality but said that we had to organize for tomorrow's return to the airport as we could be gone before the cock crows.

The next morning, Jo and I dashed into the dining room early, swallowed a cup of tea, and grabbed some bread rolls that we ate in the car on our way back to the airport.

Our return hour-flight to England, literally a sweep over the sea, went quickly. The rest of the days I had off-duty were spent at the BOAC sports club playing tennis or squash with other crew members. Then one evening, when I arrived home, Jo was busy in the kitchen making pasta, so I popped in to greet her before going to my room. In the course of a conversation, she mentioned, "I've heard stories from friends about the Prospect of Whitby in Wapping. It's a famous historic pub on the banks of the Thames. Would you care to go see it?"

"That would be wonderful. Let's visit tomorrow, as there are only a few days left before our next take-off." So, we made a spontaneous decision.

On a freezing November morning, with omnipresent steely skies stretching from one end of the heavens to the other, Jo and I caught the tube to Wapping. Nothing could deter us from taking our jaunt to see one of the oldest pub on the Thames. Outside the station we were confronted by a quiet open road with buildings on either side. Stopping a passer-by, we asked for directions to the Prospect of Whitby. He was a local in the area and directed us to the pub. We strolled along Wapping High Street to a road signposted Wapping Wall and found the pub.

Fascinated by its exterior, we stood for a few minutes absorbed in the traditional appearance of the building. "King Henry VIII was a frequent visitor to this famous old pub, dating back to the 1520s," Jo said, reading out information on a gold plaque on the pub wall.

Staring at the façade, I felt like walking through the front door we would step into a bygone era, another dimension.

The plaque showing past visitors

As we walked in, Jo remarked on the pewter-topped long bar counter built on barrels. The upright timber pillars appeared to be parts of a ship's mast. She had read about the features as she had received a leaflet on the tavern a few days before our visit.

Glancing around, I nudged her. "You can imagine bawdy, boisterous pirates with large brass rings dangling from their ears, lounging over the tables. Roaring with laughter and swigging tankards of ale while planning their next felonious feats."

Jo and I surveyed our surroundings with wonder, making our way from room to room, fascinated by the old photographs and fixtures, before venturing onto a small balcony. Peering over the side at the river below, I noticed a tall, thick, wooden block standing upright next to the tavern wall. It was anchored in the ground, with a protruding arm and a hangman's noose swinging on the end of the arm above the water. One of the pub staff wandered over and pointing to the fixture told us it was here, in days of yore, that pirates, smugglers and mutineers met their fate. They were hanged over the water, then their bodies thrown into the river, and we were looking at the original noose. I found the story about the first ever noose hard to believe, though the rest of his tale seemed credible. Jo and I glanced at each other and shuddered.

The hangman's noose

The tide is out showing the pole with noose

Despite the miserable misty weather, we went to the front terrace, huddled around a table and ordered hot coffee from a server lurking nearby. The mist thickened as we sat sipping our coffee, listening to the lapping of the waves, and taking in the panoramic view of the Thames. Ships horns resounded through the approaching fog as the vessels plied the river.

"Jo, it's time we left before it gets any darker."

"I suppose we'll come back another time, in better weather, and meet up with some of our friends."

On our way out, we were careful to traipse over the 400-year-old flagstone floor. Not that our light steps would have done any damage after hundreds of years of use by the heavy boots of villains and smugglers.

When I stepped out into the street, I felt as though I had been in a time capsule and had to shake my head to return to reality. By now the foggy mist was clamping down faster than expected and swallowing up the buildings and cars around us. So we hurried to the train station while we could still see it.

Returning back to the flat, exhausted, we were satisfied with the day's venture. Mabel had a delicious high-tea set on the table in her cozy lounge. The red glowing embers from the fire exuded warmth throughout the entire room. And in this homely environment, we dined and chatted for a couple of hours before retiring for the night.

FLYING SOUTH

⁂

One of the first things for pilots to check anywhere in Africa was the animal situation for landing or taking off. It was January 24, 1953, sometime around noon as the Comet circled Entebbe Airport in Uganda. Before we could make our final touchdown many of the locals had to run onto the runway waving their arms frantically to clear it of loitering goats, a scattering of zebras, antelopes, or other stragglers.

On our return from a trip to Johannesburg, we were spending two days in Entebbe, a town on the shores of Lake Victoria, that's quieter, greener and more peaceful than the bustle of Kampala. BOAC crews were always accommodated at the pleasant Lake Victoria Hotel, a beautiful old colonial-style compound only a 15-minute drive from the airport. The hotel terrace overlooked the immense lake. I was always pleased to enter my spacious, clean, comfortable room after a flight. Since it was late afternoon, I quickly showered and changed before meeting the rest of my crew on the terrace. We listened to a Ugandan dance and tribal marimba band while watching the red and orange hues of the setting sun melting on the water. We dined on a magnificent buffet cuisine with local delicacies and drinks. Late into the evening we were invited to join in the traditional dancing, then continued listening to the excellent marimba band. It had a diverse repertoire of modern and traditional music that filled the air with harmonious sounds. The weather this time of the year, summer in the southern hemisphere, was perfect.

The following morning some crew members who enjoyed cricket joined the local residents of the cricket club for a morning's entertainment, while some decided to take a cruise around Lake Victoria on the African Queen and others went for a dip in the lake's cooling waters. On this occasion there were two crews staying at the hotel. Several of us were feeling the effect of the heat and humidity and headed for the rocky, sandy embankment of the lake.

Entebbe – African Queen taking passengers around the lake

Entebbe – Ferry carrying people to the African Queen

Captain Brown and Patricia Crawley were both from the outgoing crew and captain Bainbridge and I were on the same crew. We all changed into our swimwear and went to the shores of the lake, with the idea of plunging into the water.

Entebbe – Cpt Brown, Pat Crawley and Elaine

Entebbe – Cpt Brown, Cpt Bainbridge and Pat Crawley

We had been prewarned that there were many species of African turtles, as well as a large population of Nile crocodiles and to be vigilant and sensible. While generally shy, crocs are also creatures of opportunity. Sitting at the water's edge, soaking up the sun, nattering and scanning the immenseness of the scene in front, initially none of us noticed a large snout followed by two beady eyes gliding through the water several feet away from the bank. Captain Brown was the first to catch a glimpse of the movement and ripples in the water. It was an enormous crocodile surveying his kingdom. We quietly moved away from the shore, changing our minds about taking a swim. Returning to the hotel, we laughed off this death-defying incident and spent the rest of our leisure time before lunch competing with one another on the tennis courts.

Mid-afternoon I received a call from Eric Wells, an old friend, whom Audrey and I met on the ship and spent time with on the Kenyan coast. He was the chief veterinary officer for Uganda and currently living in Entebbe. He said he would meet me and time permitting take me on a journey through Entebbe's history. After visiting various historical sites, we stopped at one of the oldest Catholic Churches in Uganda. They referred to it as the Church of Bugonga, an elegant old redbrick building, once used, entrance with thatched roof, standing in the surroundings, looked neglected. It was no longer used for the entrance and exit of the parishioners. The lovely gardens were now patchy and overgrown with long-dead yellowing grass. To compensate for the unkempt exterior, a choir practice must have been taking place inside the church. The low melodious chant of local voices wafted through the sultry air. Their tones echoed loud and clear in a joyous religious chorus. The voices of the men and women symphonizing, creating beautiful harmony.

I took several photographs of Eric in the churchyard before he drove back to his home. When we arrived, we cooked and enjoyed a hearty meal before returning to the Lake Victoria Hotel. Sadly, that was my last encounter with Eric. Another ship that passed in the night.

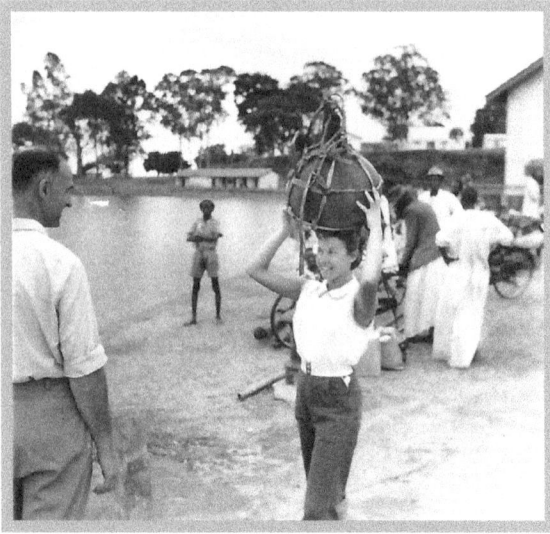

Entebbe – Elaine trying the art of carrying a calabash on her head

Entebbe – Radio officer McMahon and Elaine holding a calabash used for water

IN THE TROPICS

⌇

I t was another very dark, frosty night on the 6th of February
when I was rostered to Colombo. Before departure, I went
into our operations pre-flight building to read the essential
notes, check the passenger list and ensure there were no special
requirements for any of the travelers. I noticed we were carrying
two VIPs as far as Rome, Sir Alexander Korda, the movie
magnate, and the British actor Laurence Harvey.

I strolled across the tarmac to board the Comet G-ALYU,
where I joined my steward Bob Humphries. He was already on
board, sorting his bar and preparing sandwiches to be served
after take-off to our 33 passengers. As we were leaving just after
midnight hopefully our travelers would be sleepy after ascent, so
we may have a leisurely few hours to Rome. The crew on this
flight to the east was captain Blatchford, with first officer Gray,
engineer officer Williams and radio officer Nixon.

Once all the passengers had boarded and buckled in their
seats, I stood at the entrance door awaiting the captain. The
other crew members were already at their posts in the cockpit.
As captain Blatchford approached the aircraft, I noticed he
walked with long smooth strides covering a great deal of ground.
Even from a distance, he struck me as being a composed,

dignified character. He walked up the steps and stooped to enter the vestibule. It was hard not to notice that he was incredibly good-looking, tall and rather alluring, his dark hair curled over the sides of his cap and his uniform added to the appeal. Our eyes met. He had striking hazel eyes and a kind, classically handsome face, the type you can linger on. We conveyed the usual formal greeting, but there was something magnetic that passed between us. He gave me his overcoat to hang in the cupboard, with a hint of a smile, then strode down the aisle to meet with his colleagues in the cockpit. My steward came back and assisted me in closing the aircraft door. Then, he reported to the captain that the ship's papers were on board and everything was secure.

Our VIP, Sir Alexander Korda, was seated next to a window on the port side of the main cabin. In contrast, Laurence Harvey sat in the more private forward cabin, where he had the area that normally seats four people, to himself. A family of four occupied the table opposite and were unconcerned that a celebrity was sitting nearby.

Once airborne, I walked through the cabin, chatting to the passengers, and taking orders for drinks. Very few were interested as they preferred to recline their seats and doze once the cabin lights were dimmed. So, I walked into the galley to find out if Bob needed any help. He handed me a tray with a plate of ham sandwiches and cups of tea to take to the cockpit. I placed the tray on the end of the radio officer's table. Captain Blatchford glanced up at me, with a lingering smile. "Miss Baker, if Laurence Harvey is awake, would you please ask him to sign my daughter's autograph book," and handed it to me.

"Certainly, sir."

"Oh, and by the way, he's making the movie *Romeo and Juliet* in Rome. I don't want any Romeo-ing and Juliet-ing on the aircraft," he teased, winking, as I closed the door. Captain Blatchford had a subtle sense of humor and easy way about him, and I sensed I would enjoy working with him.

Laurence Harvey appeared deep in thought, gazing out the window at the airy kingdom of clouds. I leaned over and said quietly, "Excuse me, Mr Harvey." He glanced up and smiled. "Please, may I trouble you to sign this autograph book for the captain's daughter?" I returned the smile and handed it to him.

"With pleasure." He immediately opened his briefcase from which he took out a pen and several photographs.

"And please, would you mind signing a picture for me."

"Not at all! This flight is a wonderful experience." He handed the book back to me with a press photo placed inside for the captain. Then he presented me with an inscribed picture which read, 'To Elaine, in appreciation and the great joy of my first COMET flight. Best wishes to you. Sincerely, Laurence Harvey.'

"Thank you, Mr Harvey. It's a lovely photo. May I get you something to eat or drink?"

"Not at this moment," he gave me another beguiling smile and turned back to stare at the star-spangled heavens.

The photo from Laurence Harvey

EVERYONE in the main cabin appeared settled, some chatting with one another, others smoking or dozing. Only a few passengers accepted the offer of tea or coffee. It was too early in the morning and they wanted to rest.

We arrived at Ciampino airport in Rome at two in the morning. The producer Sir Alexander Korda and the actor Laurence Harvey, along with two other passengers, disembarked and dawdled over to immigration as they were half asleep. I chaperoned the remainder of the drowsy travelers to the restaurant while our aircraft was being refueled. When we left Rome a little before 4am, we had a full load of 36 passengers. The flight continued on to Beirut, then Bahrain where we, the crew, had three rest days to ourselves.

After the break, our captain took over Comet G-ALYS from Bahrain. Then continued via Karachi and Bombay, reaching Ratmalana Airport in Colombo, Sri Lanka, on the 10th of February at dawn, just as the sun was beginning to spread its glorious bright rays across the sky. A full 24 hours of leisure in this tropical island paradise, the so-called Pearl of the Indian Ocean, was going to be enjoyed by us all.

Whizzing through immigration and customs, we found the crew bus waiting outside the building. I wondered how long it would take this battered old vehicle to reach our hotel, 20 miles from the airport. Yet the drive was impressive as our bus jostled along rutted pot-holed roads and narrow stretches of tarmac bordered by coconut palms.

Coastal road to the hotel

We passed a section where an Indian riding an elephant could be seen pushing a tree trunk to a pile of logs waiting to be loaded onto a truck. Our driver, noting our curiosity, paused for a moment on the roadside. So that we could see what he referred to as 'logging operations' and promised he would tell us more about it at a later date.

In the warmth of the morning sun, I heard the sound of cicada beetles hiding somewhere in the dense tropical vegetation. It was a noise so familiar to me from my days living in Africa. I was enthralled with these new scenes of sights and sounds. Finally, reaching the Mount Lavinia Hotel, tucked on a small cliff with terraces spilling down to the beach and the rocks below. The sprawling whitewashed colonial-era structure, one of the oldest hotels in Sri Lanka, had its own private beach of golden sands and you could hear the roar of the ocean just outside.

Mount Lavinia Hotel

Walking into my comfortable room, I went straight over to the large window and threw it open to the fresh, balmy sea air, soaking it all in. The vista of the sea and the spectacular palm-studded beach and coves made me think of Paul Gauguin's colorful paintings of the island of Tahiti. I didn't want to miss a moment and changed into a crisp cotton dress and strolled downstairs to meet the rest of the crew for breakfast on the sea-facing terrace. It was so peaceful listening to the lapping of the waves and watching the swaying palms in the breeze. We all felt somewhat lethargic, not having slept for several hours, but nattered while consuming our light meal.

As I nibbled away at a piece of toast, I noticed a young Singhalese man wearing a long white robe with a tatty red fez on his head. Sitting at the bottom of the veranda steps playing an eerie tune on a reed flute, he piqued my curiosity. A covered, sizeable, oval-shaped straw basket placed in front of him. Intrigued, I excused myself from the table and went to investigate.

"Excuse me, what do you have in the basket?" I asked.

"Picture, you want picture?" he queried with a Sinhalese accent.

"A picture of what? I'd like to know what is in the basket."

"I have a cobra."

"Oh, please let me see it."

He removed the lid and piped a different squeaky strain on his flute. A large, tired-looking snake wormed its way to the top, extending its body in an upright position while rising and falling. The poor reptile looked exhausted from its monotonous routine. It was repeatedly in and out of its basket for the public's entertainment. Though somewhere, sometimes I had heard in the art of snake charming that the snakes are hypnotized by the rhythmic movement of the charmer's body.

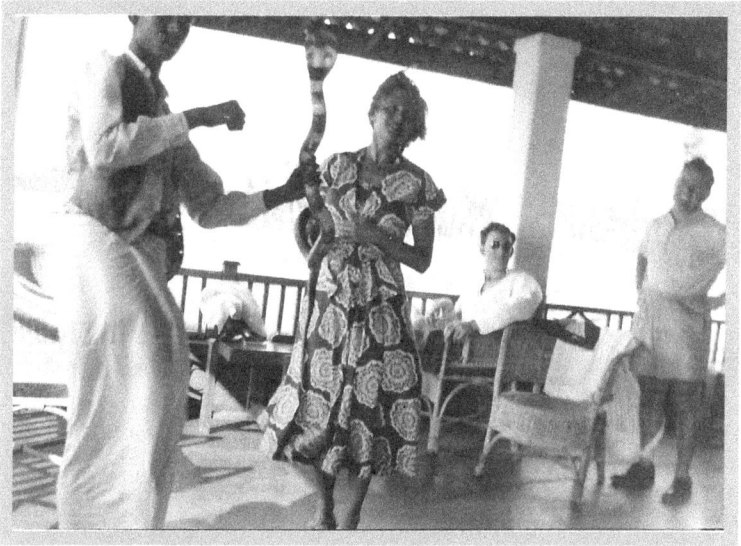

Snake charmer

Our engineer Reg Williams had followed me from the terrace carrying my camera. The snake charmer continued his hustle, asking me if I'd like a picture with the snake around my neck. I was reluctant, feeling sorry for the poor snake, and declined.

Then he said in his simple well-rehearsed English, "Not to be afraid. Vivien Leigh here last week and she had picture taken with cobra around neck."

Mr Williams glanced at me, expectantly, with a broad grin.

"Well, if Vivien Leigh could do it, so can I," I said, hoping it was the same reptile that encircled her neck. Someone at the hotel later informed us that the actress was in Colombo shooting the movie *Elephant Walk*.

The charmer played on with a piercing sound emanating from his pipe, and with one hand, lifted the cobra out of its basket of security. He positioned the cobra gently on my right arm from where it slithered onto my shoulders and draped itself around my neck. Fortunately it was a well-trained reptile.

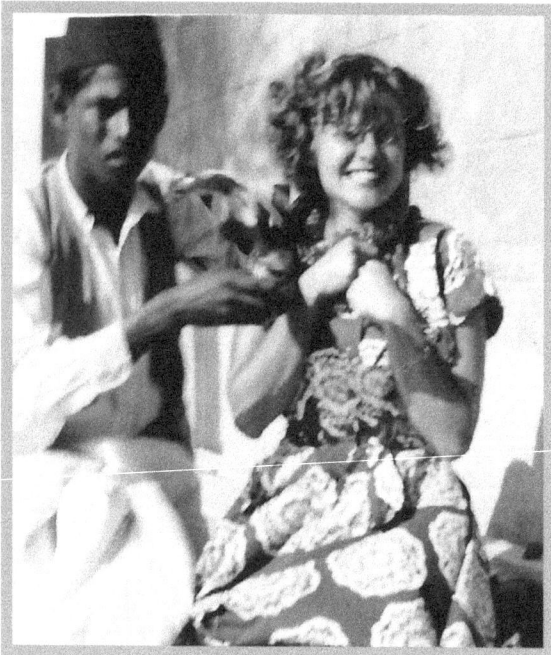

Elaine with the cobra around her neck

I looked sideways with fear and realized its flared hood was close to my face. But after a few minutes, composed, I thanked the charmer and asked him to detach the cobra from me. He extended his hand gently, removing his reptilian partner from my shoulders, to my relief. Meanwhile, Mr Williams had been busy snapping pictures and then paid the charmer 50 rupees before we returned to the terrace with our crew.

We spent the afternoon on the beach to make the most of the time on this stunningly beautiful island. Tossing our towels onto the prow, we raced one another into the water, laughing and splashing like a bunch of schoolkids. I was about to swim out into the ocean when captain Blatchford called to me as he splattered seawater over his head, "Would you like to go a walk along the beach?"

Reg, Brian and Peter

I looked over at this handsome man, who'd swept me away from the moment I laid eyes on him, in wide-eyed amazement. "Sure, I'll go," I beamed, hoping I didn't sound too eager. Leaving the crew frolicking in the waves, we walked along the shore, hand in hand, towards the far end of the sands, greeting the odd local climbing a palm to pick coconuts. It felt like the most natural thing in the world.

Spectacular rocky beach with palm trees

When we reached the end of the beach, we sat on a massive rock. Gazing out at the endless ocean, stretching as far as our eyes could see, we were absorbed in each other's company. He turned to me, smiling, his clear hazel eyes shimmering in the sunlight. "You're wonderful to be with. I love everything about you, you're natural and unpretentious."

"I'm delighted to be with you too. You know so much about life and I find your stories relating to the war interesting. People I've spoken to in the past would never recount their part played in aerial battles." He was so dashing and such great company.

I bent down to pick up a large mussel shell that was washed onto the sand with smaller variegated shells. "My father Lance Corporal William Butlin was in the South African Transvaal Scottish Division," I said. "He was taken prisoner by the Germans with thousands of others when Rommel captured Tobruk on the 21st of June 1942. He was interned in a concentration camp in Germany. It was called Annaberg. He has never talked of his experiences during his time in captivity, and always seems reluctant to do so."

"Hmm... conditions were brutal in the camps. No doubt countless ex-prisoners of war preferred to erase events that occurred from their minds."

Then Brian, as the captain asked me to call him, started telling me more about himself while I settled on the rock with the shell in my hand. "I've mentioned before I was in the RAF flying Lancasters. We had several precarious sorties, but I was fortunate enough to be one of the lucky survivors. Many of my friends lost their lives." I also found out that he was 32 years old, had married young, now divorced and had a daughter who lived with her mother.

He talked and talked, and then I told him a bit about my life. "I was very young when the war began. My sister Audrey and I went to St Anne's College, a private boarding school in South Africa. The family wanted us to be placed as far from Northern Rhodesia as possible in the event of the war in North Africa escalating further south. Apart from the school curriculum, the college held news-related classes three times a week. They did this so we were kept up to date with international current events, particularly about the war."

Brian put his arms around me and drew me closer to him. I smiled, melting into his arms, and placed my head onto his shoulder as we sat gazing at the changing mood of the sea. I looked up at him, smiling, squinting, with one hand up over my eyes to block the sun, and innocently asked, "Brian, do you think we have any kind of destiny together?"

Without answering, he held me tighter, kissing the top of my head. We spent more leisurely moments talking, enthralled with one another's stories. Pulling away, he looked into my eyes and said, "Everyone has a destiny in life."

As a slight wind blew, waving the palms surrounding the beach, Brian grabbed my hand, and we ran back towards the yacht. Splashing in the wavelets lapping onto the shore, he stopped to pick up stones and made them skip across the water. I dug my toes into the soft, golden sand and collected a few more shells while watching little sand crabs burrowing into their holes.

The afternoon passed quickly, too quickly. I was starving when we got back to the catamaran where we had left our things. The rest of the crew had vanished. Brian and I picked up our towels and wandered back to the hotel to rest and change, before meeting again on the terrace for dinner.

I lay back and soaked in a bath, washing the sand and salt from my hair. When I closed my eyes, I thought of him, our salty kisses and the wonderful few hours we had spent together. Feeling fresh and clean, I stepped out of the bathtub, dried myself, popped into my kimono and lay on the bed for half an hour. Afterwards I dressed for dinner in a summery off-the-shoulder, calf-length, pale-yellow chiffon dress with a very full skirt, nipped in at the waist. I tied a pair of dainty gold sandals to my feet and went to meet Brian.

It was a gloriously warm and balmy evening. The stars were glowing in the heavens as we sat on the terrace, dining alone at a small candle-lit table. Our colleagues were nowhere in sight. They had probably headed into town to meet with old friends. The head waiter approached wearing white gloves and an immaculate starched uniform. In a very English accent, he asked, "Can I get you something to drink?"

Brian, looking at me over the top of the wine list he was holding, asked, "Should we try a glass of white wine?"

"Why not? This is a special occasion."

"Thank you, we will each have a glass of Chardonnay," he said, glancing up at the server.

The exotic setting and the music made us feel romantic, understanding that it was a moment which would pass too fast. A band in the background was playing dreamy, wistful tunes, some reminiscent of the war years, *I'll be seeing you*, and *As Time Goes By*. Captain Blatchford took my hand, leading me onto the dance floor. Between the flowing wine and dinner, we danced for the rest of the night. The entire setting added an ambience to this enchanted evening. Midnight arrived and the music ended with *Goodnight Sweetheart*. It was difficult letting one another go, but we had to be up early the following morning to catch our bus for the airport. I was sorry to be leaving Colombo. Another magical place I will never forget.

~

THE FOLLOWING DAY, G-ALYS took off at 8.20am, soaring into a clear blue sky and leaving behind the tropical scenes of Ceylon. We landed at Bombay and Karachi to pick up passengers and refuel. Then continued on to Bahrain, where we were spending another three layover days before taking the next incoming aircraft to Rome. Early evening on the 11th of February, our captain made a gentle touchdown when landing on the Bahrain runway right on time. Then, after the usual immigration and customs formalities and the drive to town, we reached the BOAC rest-house about eight o'clock.

Inside the foyer, the crew dispersed and captain Blatchford pulled me aside. "We'll meet back here in 15 minutes to go to dinner before the dining room closes."

"All right. I won't be long... just dropping off my bags."

On my return to the lobby, I was confronted by a tall, swarthy gentleman in a dark-gray pinstriped suit, a blue shirt and gray tie. He approached me, smiling and said, "I've come to take you out."

"I beg your pardon? Thank you, but I'm waiting for the captain." He was a complete stranger and I was a little stunned by his brazenness.

"I would like you to accompany me to..."

At that moment, captain Blatchford appeared. Glaring at the intruder, he took my arm, walked me straight through the entrance hall door, and across the courtyard to the restaurant.

"You arrived at the right moment. I was feeling nervous of that man."

"Some of them believe that it is their prerogative to requisition a stewardess. By no means do they intend to be offensive," he looked at me with his penetrating hazel eyes.

A little while later we were joined by the rest of our crew in the dining room. We all felt a little tired and, after the coffee, I returned to my room and fell into bed.

~

CAPTAIN BLATCHFORD and I spent three blissful days together. We hired a car and scoured the timeless desert of this island kingdom on the Arabian Gulf. An occasional nomad market dotted the dunes, the owners selling pottery and colorful Persian carpets. I bought two small wall-hanging rugs, one for my kindly landlords Mabel and Ossie, and the other for my dear mother.

At the end of the last day, we drove into the desert to watch the sun go down, knowing that flaming-orange sunsets in this part of the world were spectacular. Stopping the vehicle on top of a dune, we encountered a tiny oasis in the foreground. Before us, the few green palms and several cacti plants encircling the small patch of water added a hint of color to the vast expanse of sand. Above us, the mesmerizing psychedelic glow of the setting sun provided the perfect backdrop to this tranquil scene.

Brian held my hand, gazing in awe at the changing colors of the huge desert sky. Entranced in a mysterious, friendly silence, finally broken in translucent darkness by the facade on the front. The evening star appeared luminously hanging in the night sky, and only then did he start the engine, and we returned to our hotel.

CITY OF DREAMS

On the 15th of February, we took over G-ALYY from the incoming crew and left Bahrain just after midnight. Winging our way to Rome, the Comet encountered an unforeseen problem and had to make unanticipated diversions to Cairo and Naples. Accepting the situation as normal routine, the passengers showed no signs of annoyance.

Our flight time between Cairo and Naples took five hours. Then the sector from Naples to Rome lasted less than an hour. So, on this part of the journey, I suggested to my travelers they kept their seatbelts fastened, as we would no sooner be in the air than we would again land.

Rome, the wildly romantic eternal city built on seven hills, carries the sort of magic that's difficult to find anywhere else in the world. The gold light, awe-inspiring architecture and art, the spectacular food, the stray cats roaming the ruins as they pleased, there was no place like Rome. I loved London, but I loved Rome more. In fact, more and more each time I passed through. G-ALYY finally touched down at Ciampino Airport in Rome at 5.15pm, in cold but clear weather. It was very disheartening to have to deal with a temperature of around 55 degrees after the wonderful warmth of the Far East. After a 20-mile drive from Ciampino to the city, the crew bus dropped off the flight crew at the Hotel Quirinale. Then took us, the cabin crew, to the Plaza Hotel. In Rome, the crew always split up, staying in different establishments. I have never understood the logic for this separation and never thought of finding out.

Once in my room, I organized my clothes, placing those for the cooler weather in the forefront of the pile. Then, as I was about to pop into a bath, the phone rang. That smooth, beautiful voice was so familiar. It was captain Blatchford. "Hello, would you care to go to a movie this evening? *Singing in the Rain* is showing at a small theater somewhere in the Plaza's vicinity, and I know you enjoy musicals."

"Oh yes! I'd love to."

"I'll meet you in about 45 minutes, which will give us time to get ready."

Although the evening was chilly, we were both enraptured by being together and felt more warm than cold. Both of us joined the queue of people gossiping while waiting to buy tickets. Inside the theatre, we settled ourselves in plush, comfortable seats, relaxing as the lights dimmed. The film kept us spellbound watching Gene Kelly's lithe movements and listening to the popular vocals. His masculine, athletic dance style was so beguiling I couldn't take my eyes off him. Debbie Reynolds, too, graceful and lovely, was an exceptional dancer.

Brian looked at me and smiled as he squeezed my hand. When the film ended, we walked out of the movie house into the darkening night, happy and carefree. We strolled along the banks of the River Tiber, aglow with lights beaming from the restaurants close to the river. It was very romantic, wandering along the embankment with the Tevere on one side, and watching the world passing by on the other. A low stone wall appeared, lining the waterside. Feeling content and unfettered, I leaped onto it and danced along the top impersonating Debbie Reynolds while singing *You Were Meant for Me*.

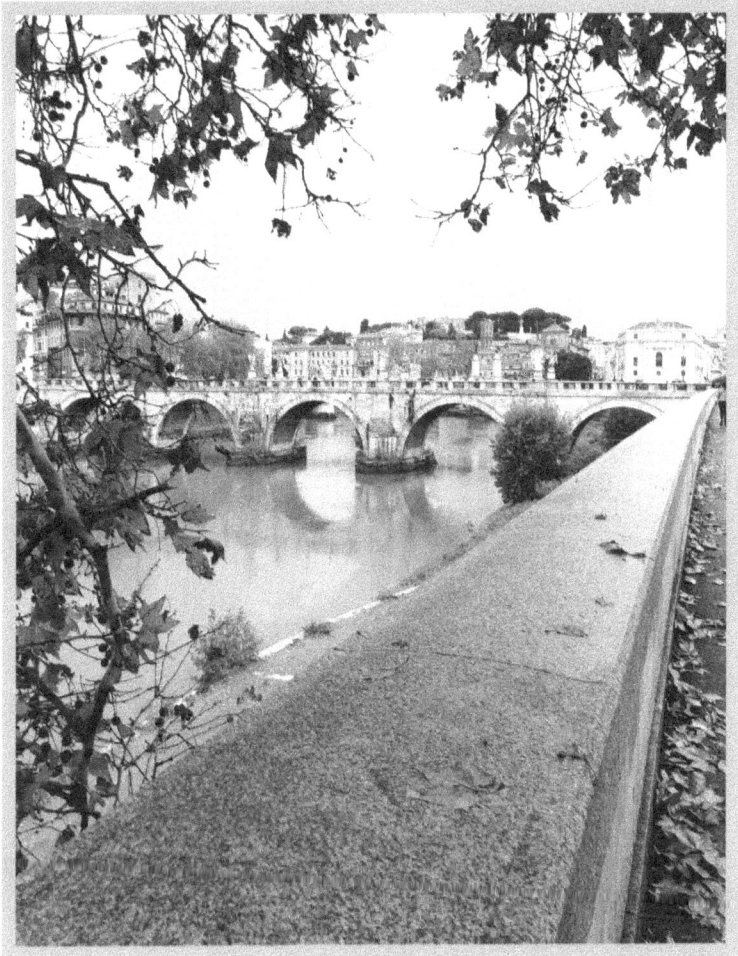

The stone wall I danced along

Brian looked up at me, laughing, then he put his hands around my waist and swung me off the wall, my full skirt swirling as I landed on the sidewalk.

It was still early evening. Outdoor tables lined the streets despite the chilly weather. Folks drinking, eating and enjoying themselves. As we passed a small, villa-lined piazza, a busker playing a guitar crooned sentimental Italian melodies with onlookers joining in. Two popular songs were *Trieste Mia* and *Te*

Voio Ben. It was captivating listening to his swarthy, melodic voice and watching his amorous movements.

On our stroll back to the hotel through the back streets, Brian and I passed an occasional vendor on a street corner selling newspapers. Some headlines published warnings of doom and gloom. Undeterred by the ominous titles, the seller still had a smile for passers-by. This made me realize no matter what events took place, Rome seemed to always be a lively and vibrant city.

Sometime around midnight, we made it back to the Plaza Hotel. Outside the entrance, Brian gave me a warm hug and kissed me. "*Buona notte, mia cara,*" he said, as he pulled away from the embrace.

"Thank you for such a wonderful evening," I answered, waving as he walked away.

"I'll call you in the morning."

∾

THE NEXT DAY after breakfast when I walked into my room, the phone was ringing, so I dashed to the bedside table to pick it up. "Ciao."

"Are you ready for a tour of Rome?" Brian asked. "I've hired a Vespa for the day."

"Wonderful! Where are you?"

"I'll pick you up outside the Plaza in ten minutes. You'll need a sweater as it might be a little chilly."

I grabbed my handbag and thick jumper and rushed downstairs to the hotel lobby. The Vespa arrived within minutes.

"Jump on and hold tight," Brian said, windswept and smiling. "We'll start at the Spanish Steps."

I put my arms around his waist and clung on as tightly as possible, never having ridden a Vespa in my life, not even a motorbike. "Guess who sat at the table next to mine at breakfast?" I spoke into his ear as we breezed along.

"I have no idea," he said.

"John Gielgud."

"Did you speak to him?"

"Only saw him briefly, then I left to meet you. He's probably here making a film in Rome."

When we reached the Piazza di Spagna, Brian found a place to park our bike for a few hours. We then strolled to the Spanish Steps, in Italian called *Salivate della Trinità dei Monti*. Brian spoke Italian reasonably well and tried to teach me names in the country's language where appropriate. It was fun walking up the 138 steps to the Trinita dei Monti Church at the top. From here we had a panoramic view of the city laid out before us, plus the fascination of people watching. Everyone seemed relaxed and happy. As we walked back down, Brian asked me if I would care for a gelato, as there were several people enjoying cones brimming with ice cream.

Spanish Steps

When we reached the end of our descent, we stopped to look at a plaque set in the wall of a house on the corner. This had been occupied by the poet John Keats. He lived there for many years and died from tuberculosis at the age of 25. Other English poets like Shelley and Byron, mesmerized by the city, lived here too and were buried, along with Keats and many famous artists and politicians, in the ancient *Cimitero Acattolico*, where hundreds of cats wandered and basked in the sunshine.

This corner house on a cobbled lane, overlooking the Spanish Steps, was classified as a museum dedicated to Keats' memory. Still, we didn't have the time to venture inside, as there was so much to be accomplished on our day's agenda. Looking at the building, I quoted, "Season of mists and mellow fruitfulness..."

Brian continued, "Close bosom-friend of the maturing sun..."

We glanced at one another and laughed, having both learned the poem *To Autumn* during our college years.

"Keats painted an imaginative picture of fall. I enjoyed his poems very much," he said.

We wandered through milling people and the maze of cobbled streets. It was interesting glancing into the windows of small shops that sold leather goods, jewelry, religious ornaments, jackets, dresses and shoes. We highlighted our favorite shopping spots, streets and communities. Winding through the impossibly narrow streets, pausing at one of the many gelato shops to try their ice creams and bought two popular flavors, one pistachio and one *dulce de leche*.

Walking out of a small alleyway, eating our ice cream, the Trevi Fountain came into view. I stood gazing, entranced at this magnificent structure, a rococo affair of mythical figures and wild horses, with rushing water around it. "It is so impressive."

"Yes, it was designed by Italian architects Nicola Salvi and Giuseppe Pannini." Brian gave me a crash course on the history behind it. "That's Oceanus in the center, the Greek god of the sea. He's riding a chariot in the shape of a shell pulled by two sea horses. One horse is calm, but the other is very restive with a wild expression in his eyes. The horses symbolize the moods of the sea."

"Gosh! It must have taken years to build," I answered, mesmerized by this marvel.

"It took about 30 years to complete, but that's another story. Anyway, if you wish to return to Rome, you must turn your back to the fountain and throw a coin over your left shoulder into the water."

I fished through my purse and produced two coins. I looked at Brian dreamily. "Maybe we'll meet again in Rome if our flight paths cross," I said, handing him a coin.

"Hopefully we will," he replied, smiling.

We turned our backs to the fountain tossing the coins over our shoulders. Then, turning around, we watched them sink to the bottom of the fountain, joining thousands of others.

Trevi Fountain

Our next stop was the Mouth of Truth. In Italian, La Bocca della Verità, which is on the portico of the Church of Santa Maria in Cosmedin. I had heard a great deal about it from colleagues on previous stopovers. In medieval times, it was an ancient Roman drain cover with an enormous face, but now it's fixed in the church's wall. Put your hand into the mouth of the mask and, provided you're a truthful being, you will retrieve your hand unscathed. At first, I was apprehensive about putting my hand into the dark gap, as I wasn't sure what might lurk inside. Anyway, we took turns thrusting our hands into the gaping mouth and retrieved them in one piece. There was so much to do and see in Rome that it would take a lifetime to become acquainted with everything it has to offer.

La Bocca della Verita

Brian loved wining and dining. As we wandered away from the Piazza Bocca della Verità, he asked, "Where would you like to have lunch?"

"I don't know. You've spent more time here than I have. Surprise me."

"All right, let's collect the Vespa and ride to Ristorante Cecilia Metella. It's in the Catacombs area along the Via Appia, and they serve superb food."

When we arrived, he parked the Vespa outside the terrace. Delighted that he had brought me to this luxurious restaurant on a low hill off the Appian Way. It was a charming place, secluded from the commotion of the city. Although the day was cool and overcast, we sat outside for lunch under a vine canopy with glimmers of the surrounding luminous-green countryside.

The waiters were very warm and chatty, assisting us with the names of the Italian dishes on the menu. On our server's recommendation, Brian had *cozze gratinate* (baked cheese-stuffed mussels) and I settled for *risotto ai frutti di mare* (seafood risotto).

After a romantic meal and bottle of Chianti, we felt mellowed in this tranquil environment and wandered through the gardens to look at the tomb of Cecilia Metella. It was built around the year 50 BC. I couldn't imagine there ever was a year dating back so far in time. As we rushed towards the Vespa, we laughed and reminisced.

"Brian, on our return, please could we stop at the Church of Bones? My friend Patsy has told me so much about it," I said, as I hopped onto the back of the scooter.

"The correct name is *Santa Maria della Concezione dei Cappuccini*."

"That's a mouthful."

Santa Maria della Concezione dei Cappiccini

When we reached the Via Veneto, close to Piazza Barberini, across the road from the Triton fountain with the alluring statue of the kneeling merman blowing into a conch shell, we found the church without difficulty. Inside the entrance, the receptionist told us that the underground crypt was divided into five chapels. As the afternoon light was fading, we made our way to the vault where Brian and I temporarily parted ways. At the entrance of the crypt, I stared with wide-eyed wonder when I saw all the walls adorned with the bones of Capuchin monks, a Franciscan order of friars established in the 16th century, who lived in piety and self-denial. It was surreal, even the chandeliers in the ossuary were made from skulls and different bones of the anatomy. There are approximately 4,000 bones in the church vault. I carefully drifted from one chapel to another as they were dimly lit by small, fluorescent lamps.

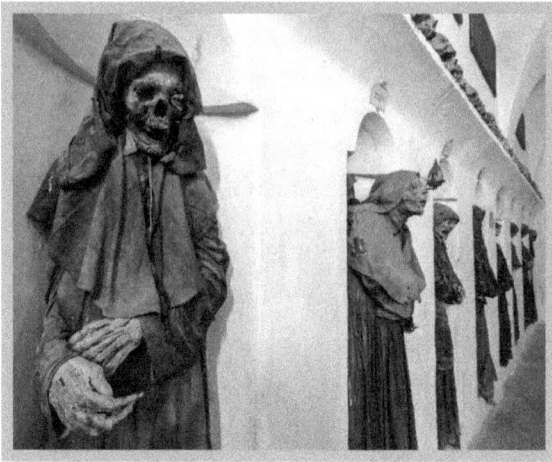

Memento Mori in the crypt of the Capuchins

Cemetery of the Capuchins

My friend had been perusing the many chambers. After examining the complete skeleton of a Capuchin monk dressed in a long dark hooded robe, we ran into each other. "Have you seen enough? This mausoleum is closing in a few minutes, so we had better leave."

"I wouldn't want to be locked in here for the night."

"There's a sign on that wall which reads, 'What you are now, we once were, what we are now, you shall be," he said, pointing to it and looking at me with a sideways smile.

We turned our backs on the gloomy past and walked out into the breezy evening to collect the Vespa. When Brian dropped me off at The Plaza, I said, "Such an exciting day. I'll never forget our time together in Rome. Moving from place to place on a Vespa has been such great fun."

"I'm so glad. I wasn't sure whether you would approve of our transport. And don't forget the party at La Bibliothèque tonight." Then he whizzed off to return the scooter.

That evening, the outgoing crew and our crew met at La Bibliothèque. I noticed on entering the club, the eye-catching display bottles of different wines from Italy lining the walls. I now understood why it was well known among airline crews for its atmosphere and popularity. The premier restaurant was in a large cellar, with small subsidiary cellars opening off it. There was an excellent band in the central cellar, combined with a small circular dancefloor. Those who wished to dine or drink with heady ambiance could retreat to the smaller cubicles, where the music was more subdued for conversation.

Patsy and Joanna

Elaine, Patsy and Joanna

Captain Peers from the outgoing flight, my friend Patsy, the stewardess on his crew, and Brian and I chose a subsidiary cellar for dinner. The rest of our colleagues preferred a large table in the central area, joining members from other international airlines. The band was playing *April in Portugal*. Several couples on the floor were dancing slowly in a tight embrace.

A server arrived and with usual grace said, "*Che cosa prendete?*"

Captain Peers answered for us all in English, "Spaghetti Bolognese, a bottle of Chianti, and a large jug of water, *per favore.*"

When the piping-hot pasta arrived, our server poured the first glasses of Chianti. Thirsty, I immediately swallowed half my glass of wine and refilled it.

"Now, child, Chianti must be sipped with concentration," captain Peers said to me, smiling.

Between the dining and drinking, Brian and I danced closely, oblivious of our surroundings and everyone else around us. Later in the evening, as we were all sitting socializing in our private enclosure, the violinist appeared. He serenaded us with his instrument and his beautiful tenor voice. We left La Bibliothèque around midnight, happily stumbling into the Roman night after our wonderful evening.

Back in my hotel room, I was about to put out the light when the phone rang. Only one word was spoken, "Goodnight" and Brian rang off. He did not have to say anymore.

∿

THE FOLLOWING afternoon we left Rome for London on G-ALYU in drab, cold weather. At the end of the runway, our aircraft soared into the air, climbing through thick gray clouds. There was a certain amount of buffeting as we ascended higher into the sky, finally breaking away from the dark cotton-wool clouds. The No Smoking lights went off, but the Fasten Seatbelts sign remained until we leveled off and reached stable air. Leaving my seat, I made the customary round of handing out complementary cigarettes. Making sure the passengers were comfortable, I let them know that light refreshments were on the way. After the tea service, followed by early drinks for the passengers requesting an aperitif, we prepared for our landing in London. The two hours 45 minutes quickly passed. It always did, with two cabin crew to render first-class service to the travelers.

Moments later captain Blatchford and I were standing on Hounslow station waiting for the train. Our 'crew' baggage labels, written in bold black on a white background, were very prominent. A kindly little old lady in a dusty black straw hat, trimmed with flowers around the crown, approached us, "Excuse me, my dears, but the train for Crewe has just left."

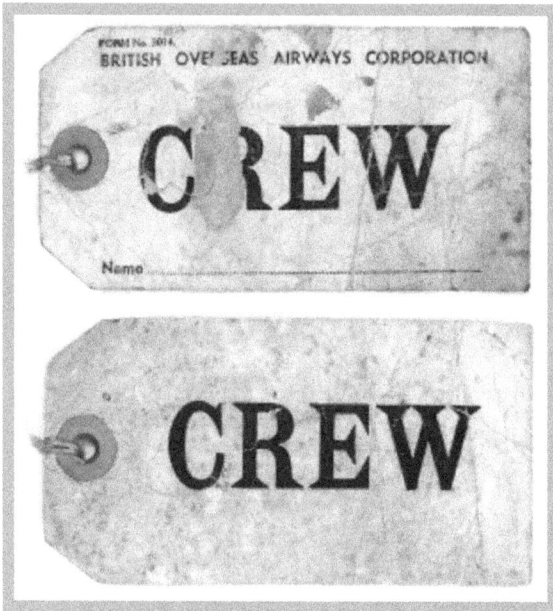

Crew baggage labels

Brian and I glanced at one another, smiling, thanked her for her concern. Then graciously told her we were actually going to South Kensington on this occasion.

"Oh, I am so sorry for intervening," she said and disappeared into the lingering crowd. The train arrived. It was almost full, but we found a couple of seats together. People kept staring at us. We seemed to be the focus of all eyes in our smart blue uniforms, as aircrew seldom used public transport. However, captain Blatchford had left his car at home.

When we arrived at South Ken station, we both slipped into a nearby small cafe for coffee and cake, spending a convivial hour together before parting. Then he walked with me to the corner of Old Brompton Road.

Holding my hand, Brian looked into my eyes and said, "I want to be with you all the time. I spend hours wondering when we will again fly together."

"I think of you continuously, wondering what fate has in store." I adored every moment with him and thought I could easily imagine a future together.

Facing me, he gave me a kiss, clutched my hand, and then turned and walked back to the tube station without a rearward glance.

After our last supper in Rome, Brian and I remained in touch. I was constantly hoping to see him somewhere en route, often thinking about him, but sadly our paths never crossed again.

A WANDERING LIFE

I had covered many miles over the past months and flown to many places. London-Colombo, London-Johannesburg, London-Singapore, London-Tokyo with layovers at various venues on different routes. Flying became a way of life, a diverse life with opportunities to visit many regions and experience cultures beyond most people's reach and wildest dreams.

On one of my off-duty nights, I was sitting on the sofa in our cozy lounge, with the curtains drawn to shut out the scene of the misty cold and rain beating on the windowpanes. While paging through my diary and book of notes, I reminisced about various incidents that have occurred at different times en route. But first, there was an entry about my visit to the House of Commons with my old flatmates from Cromwell Road.

When I was staying at 94A waiting my contract to arrive from BOAC, Pam, Paddy and Veronica suggested we attend a session in the House of Commons. Being an admirer of Prime Minister Winston Churchill, I thought this would be the perfect opportunity to see him in the flesh. I was very young when Britain stood alone against the might of the German Third Reich. As I recall, Churchill was a great orator. He kept people together with words that let them know democracy would rule victorious over dictatorship. Surely the world has not forgotten.

On a chosen day, we arrived at Parliament and joined, to our delight, not too long a queue. At the entrance, we had to open our handbags, which an attendant examined. And a security guard scrutinized us before he escorted us into the visitors' gallery. The gallery stretches around the top of the House of Commons.

It's in this vast arena where all the debates and business of running the country are conducted. And it's the main parliamentary forum for political battle. As we entered the gallery, we were cautioned by the usher that cameras were not allowed. Fortunately, we had not contemplated bringing one along.

Yet there is always someone who will disobey the rules. One visitor sitting in the row of seats behind us snuck his camera in somehow under his coat. As he attempted to take a picture, quick as a flash, a burly police officer leaned over the back of his seat. He grabbed the individual by the collar of his coat, hauled him out of his pew, confiscated his camera, and ejected him from the gallery.

Sitting in my seat, I glanced around the extensive surrounds. Down below on either side of the Speaker's Throne sat the government members and the Opposition. Immediately in front of them is a carved and polished table on which lies, in its appointed place, the Mace. Without the Mace in position, each House cannot sit and debate. This is borne into the arena at the beginning of each session of Parliament. It is positioned on an ornate table dividing the two sides of the House.

The Prime Minister, the Chancellor of the Exchequer and other cabinet ministers sit in the front row of seats to the left of the Speaker's throne. The Speaker's throne is elevated above the floor of the House. Behind them are the members representing every country in the United Kingdom. The same formation follows on the Opposition benches with the Leader of the Opposition and his Shadow Cabinet alongside him. And the representatives of their countries seated behind them.

We stared down riveted as Winston Churchill strolled into the House. He bowed to the Speaker, and sat in the Prime Minister's seat on the bench opposite the leader of the Opposition. This moving event left us spellbound.

On leaving the Houses of Parliament, we leisurely strolled along the sidewalks, chatting. Attending this session in the House of Commons was a highlight in our lives. We had seen in the flesh this great man of destiny, Winston Churchill.

Reading on, I got caught up in memories of my training last September, when our class was sent for three days to a destination referred to as Meadowbank. This was for instruction on air/sea rescue, and jungle, desert, and Arctic survival:

Our instructor was a delightful, amicable gentleman of medium height. With salt-and-pepper hair and displaying a manicured moustache on his top lip, he was always immaculate in a shirt, tie, tweed jacket and slacks. He presented himself as Mr Parrot, so one of the males in our group introduced "Polly" between Mr and Parrot. So, for the entire course, he was known as Mr Polly Parrot.

Our first day of training at Meadowbank was spent on air/sea rescue, then after that, we were taught how to survive in the jungle and other outposts. It was essential to know which plants we could exist on in the bush as many were poisonous.

Mr Parrot stated that if one was fortunate enough to find monkeys or gorillas eating certain plants, they would be safe for us to consume. Solid food included not only birds and animals but frogs, lizards, ants and grubs — split and broil over an open fire. Crickets, remove legs and wings then roast. Ugh. The list was very long and I decided I had no intentions of landing in the jungle. God willing.

*Daphne, Wynn, Dorothy, Myrna, Mr Parrot, Jo, Mary, Elaine
and 3 training stewards*

REGARDLESS, *all the lectures were interesting and enlightening. We were advised to carry in our handbags at all times a compass, a penknife and a magnifying glass. Three useful items for different reasons made known to us. The penknife was for cutting anything. The magnifying glass will help in making fires in the desert or the wild. And the compass is to assist us in finding direction if lost in outlandish places. At the end of the course, we expressed our gratitude to Mr Parrot, as we all felt competent enough to deal with any situation should a mishap ever occur.*

TURNING the page of my notebook, I reread an incident that took place in the Far East during an interlude in Bangkok: *It was one of those glorious mornings when the sun was gleaming, and the crew decided that the swimming pool was the best place to be. So we gathered our bathing items together and made our way to the Royal Bangkok Sports Club.*

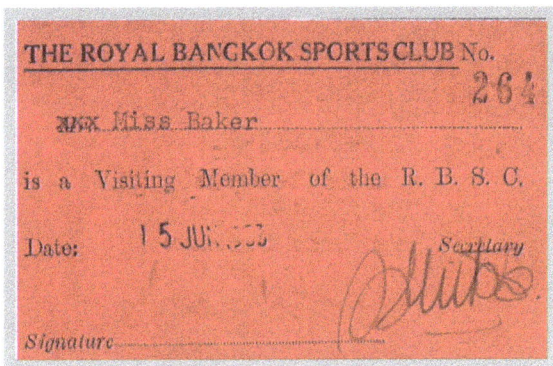

THE ROYAL BANGKOK SPORTS CLUB No.

264

xxx Miss Baker

is a Visiting Member of the R. B. S. C.

Date: 15 JUN ...

Secretary

Signature

Royal Bangkok Sports Club Membership Card

On every flight, I always carry a novel to read whenever there is a quiet moment during a stopover. On this occasion, I was absorbed with the story *A Many-Splendored Thing*, an autobiography by Han Suyin. It's about post-war Asia shaken by revolutionary changes. In the story, Han Suyin, a Chinese medical doctor, falls in love with Mark Elliot, a foreign correspondent reporting on the war. We were throwing our swimming towels into the area surrounding the pool, and I had no desire to place my book anywhere that could be doused with water.

At the opposite end of the pool, I noticed a gentleman luxuriating in the sunshine on a chaise lounge, with a table beside him. There were, in fact, only a couple of tables scattered in the area, so I sauntered over and asked him if he would allow me to leave my book on the table. He was so affable and smiling, replied. "What are you reading?" I handed the novel to him, which he scrutinized, then said, "I was the reporter who was with Mark, the actual Mark Elliot in the story, when he was killed. Mark was my close friend."

His reply stunned me, and I asked him to please tell me more once I had completed a promised swimming race with my crew. Reaching the end of the pool after the race, it was disappointing to find the chaise lounge vacant. Though a note lay on the cover of the novel which read, "Go well; sorry I couldn't wait but duty calls." It was unfortunate not having further communication with this interesting and charming journalist.

Towards midday, the weather became somewhat sultry, so we, our crew, returned to the hotel for an afternoon siesta. We meandered along New Road, the main street of the city. Chinese shops, on both sides of this thoroughfare, were interspersed with a few European-looking stores. As I walked along the sidewalk, I noticed some shop windows showcased with mandarin-style dresses. I longed to own one. Parting from the crew, I went into a store, and for a few minutes stood gazing at the racks filled with colorful Chinese garments. Pyjamas, silk kimonos, short and long brocade cheongsams, plus many other items to use as accessories for the robes, all appealed to me.

After browsing for almost half an hour, I chose a beautiful creamy-gold brocade cheongsam, plus a pale-blue silk kimono with a gold dragon embroidered down the back and a pair of matching mules. The blue kimono and slippers, I always carry with me on all my trips.

Elaine in a Cheongsam

When I left the shop, the captain and the remaining crew members had disappeared, apart from our engineer who waited to accompany me to the hotel. The others had decided to side-track through the markets.

Here, Saki sets, rice spoons, chopsticks, Chinese bowls and, again, silk kimonos could be bought at competitive prices.

Ken, our engineer, and I returned to the hotel and dropped off our purchases in our rooms. He suggested that we visit a Wat in preference to wasting away the afternoon catnapping. We walked into the street, hailed a samlor, a part rickshaw, part bicycle, drawn by a Thai local. And he explained to him we wished to go to the river to take a boat to Wat Arun. He understood our request, so we climbed in and settled back to enjoy the ride.

We soon found ourselves on the banks of a canal. Here, we disembarked from the samlor and wandered along the shore while viewing the surroundings then approached the owner of a sampan tied to a rampart. His face could barely be seen as he was wearing a large straw hat resembling a lampshade to cover him from the heat of the sun. The man was perched at the stern of his boat, playing with a ball of string. We asked him if he would take us near Wat Arun, generally known as Temple Dawn. It's very famous and a "must see" when in Bangkok. He agreed for a small fee.

Making ourselves comfortable in his vessel, drifting along the river, we passed small boats plying in both directions. Some filled to capacity with all sorts of goods. Then finally converging on a floating market, where Thai families were doing their shopping. We were not inclined to buy anything, so asked the boatman to continue paddling to Temple Dawn.

Temple Dawn in Bangkok

Wat Arun is indescribable. It is a magnificent Buddhist temple on the west bank of the Chao Phraya River. It is famous for its decorative beauty. Referring to a leaflet I collected from the hotel, the central tower of the Wat rises over 200 feet. It has carved roofs adorned with red, black and green tiles.

It would take several hours to view this temple on foot and allude to the galleries featuring the rows of legendary warlike heroes. But as there was little time for the short venture, we remained in the comfort of the skiff. Ken and I absorbed as much as possible from our riverside view and noted that the imposing spire on Wat Arun rises, over 70 meters high, on Wat Arun. It's adorned with tiny pieces of colored glass and porcelain, positioned into intricate patterns. I must visit this temple next time I'm in Bangkok, I thought to myself. Ken then asked our captain to turn the boat around and head back to the point of embarkation. So he maneuvered us to our starting spot where we paid the requested fee and thanked him for his help. He put his hands together, giving a little bow before we turned our backs and ambled away.

Gateway to the chapel of Sleeping Buddha in Bangkok

Although it was a fair distance to the hotel, Ken and I returned via the marketplace. The booths were buzzing with activity displaying varieties of wares on the counters. In the food line there were sausages, dried and smoked bats, spices, rice, eggs, fish, and many other things, none of which we fancied. In other shops, there was junk of every description. Old records, incense batons, cameras with or without lenses, electric bulbs and broken pieces of china. Semi-precious stones, such as zircons and sapphires could be bought at one stand. And the owner would set them for you in a design of your own choice.

We continued roaming through this conglomeration of anything and everything. In one booth, men were sitting cross-legged on the floor, placing bets on exquisite, fairy-like fish. My friend Ken, told me they were Siamese fighting fish, known as Betta fish.

I had never seen or heard about this sport and was entranced, but saddened, at these beautiful marine creatures fighting to kill. Besides their gorgeous coloring of corn-starch blues and various reds and oranges, they have a graceful movement in the water. It distressed me to think that only one Betta fish would survive the ordeal. Ken lured me away. Leaving the gamblers shrieking as they threw coins onto their mats and placed their bets on these tiny thalassic gladiators.

Back in the street trolleys rumbled past us, taxis tooted their horns and cars glided along. Some samlors drawn by Chinese locals were loaded

with one or two passengers and their luggage. They wormed their way
through the melee of traffic.

We were both pleased to get back to the tranquility of our hotel. At
the desk the receptionist handed me my room key, and knowing of my
interest in historical places, she told me about Kanchanaburi War
Cemetery. It lies 150 miles south of Bangkok. It is maintained by the
Siam government and contains over 6,000 graves. It's the burial place of
British, Dutch, Australian and Indian soldiers. They perished while
constructing the famous Bridge on the River Kwai for the Japanese. I
told her I would love to go to see the cemetery, but, on this occasion, there
was no time. I'd definitely make it a priority to visit Kanchanaburi on
my next stopover in Bangkok.

AS I TURNED the page of my notebook, I heard a gentle tap on
the door. Mabel pushed it open, saying, "May I come in? I've
brought you a cup of tea and sandwiches as we're having a break
for a few minutes before continuing to develop some portraits."
They were always inundated with work being well-known as
excellent portrait photographers.

"You are wonderful landlords and appreciated. Thank you
very much. I've just been reading through my diary and hadn't
thought about tea. Please sit down for a few minutes."

Mabel placed the tea and sandwiches on a small table and sat
back in a large, easy chair beside the fireplace.

"Do you remember last week I was invited out by a
gentleman called Shannon Stocks. The very handsome dress
designer from Australia, who I met on one of my recent flights?"

She laughed and said, "Yes, but you've mentioned nothing
about your jaunt with him."

"Well, before we went to the show *The Boyfriend*, he took me
to the Royal Airforce Club for a cocktail. Shannon and I were
having an in-depth conversation about wartime flying, when a
tall, slim, fashionable woman glided towards the bar with her
companion. All heads turned. I digressed from our airplane

chatter. Gazing at her glamorous outfit, I asked him if he would consider designing a dress for me, but he could decide whether it be for evening or daywear."

"What was his answer?" Mabel asked, fascinated and smiling.

"He said, 'Why should I design a dress for you to attract other people?' Well, his reply took the wind right out of my sails."

"Hmm, sounds as though there might've been a hint of jealousy. I wonder if he will contact you again," Mabel joked.

"Don't know, but we spent an enjoyable evening together."

Mabel stood up. "I must get back to the studio. Drink your tea while it's still hot," she said and walked out, closing the door behind her to keep the warmth in the room.

I drank the tea and nibbled the sandwiches and continued leafing through the pages of notes, these memories of places and people, vividly flickered through my mind like old home movies.

On one occasion, Win Min Than, a beautiful young lady from Eurasia, was on the final sector of a flight I made to Colombo. When the crew noticed her, she was invited onto the flight deck to be shown the layout of the instrument panels. Passengers are always elated when asked into the cockpit where they can view the controls of the aircraft. Not that Win Min Than was particularly interested, but it was more the fact that the crew had the attention of a pretty girl for a short time.

During the interlude, I continued attending to the passengers. Finally, the captain summoned me to escort the lady back to her seat. She chit chatted with both me and the steward as we passed through the galley. She told us she had been selected from hundreds of hopeful applicants to appear in a movie, playing opposite Gregory Peck. And had now come to Colombo to take part in the film Purple Plain.

During our conversation, she mentioned that she had a very jealous husband. He had instructed her to eat plenty of garlic throughout her time spent in close proximity with her leading man. Inwardly, the steward and I found this amusing. When she disembarked at Ratmalana Airport, Ceylon, I wished her lots of good fortune and said that I looked forward to seeing the movie in the new year.

~

CLOSING THE DIARY AND NOTEBOOK, I placed them on a small adjacent table, got up and looked out the window to check the weather. The grandmother clock in the corner chimed five o'clock. Parting the curtains, I looked at the watery outside world. The rain had not subsided. I thought, it's so cold and dark, and the clouds are still hanging like a thick blanket hiding the skies. Staring out, I was intrigued with the sea of umbrellas fighting the glass-like drops falling from the sky. For a few moments, I watched humanity hurrying by. It amazed me how, no matter what the weather, folks were always out and about enjoying their lives. Time controls their destiny, their every move. The journeys we take in our daily lives lead us all down our varied paths. Some of us write our own destiny, making sure we end up where we visualize is best. But for others, destiny is something they follow and nothing can change it.

I let the curtain close as if I had watched a final stage performance.

Walking back to the sofa, I realized that I had a thrilling job where I was always on the go, surrounded by people. Yet, for me, it was my passion to break through blankets of clouds that suffocate the metropolis. Soaring in that graceful metallic vessel with swept-back wings, and carrying me to unimaginable places in the world. When I am flying, I feel like a bird floating as it catches the wind, drifting over cities near and far. I cherish everything about it, the fascinating places I visit, the interesting people I meet, but every time I fly, I just feel free.

Much have I travelled in the realms of gold, And many goodly states and kingdoms seen...

— JOHN KEATES

BRIGHT LIGHTS, BIG CITY

ate afternoon on Monday, 2nd of November 1953, I
received a call from the head of Comet's catering
section. He instructed me to report to the office in
uniform on Wednesday morning at 10 o'clock. I didn't
understand what this summons meant or what it was about. It
sounded serious. Had I been written up in a bad report for some
minor infraction, maybe a passenger complained about
something? The fear and panic hit me like a clap of thunder. I
immediately thought I was going to have my wings clipped and
fired. One always imagines the worst, if not enlightened.

Elaine in BOAC uniform

Feeling low and a little fearful, not knowing my imminent fate, the following morning, I called my close friend Patsy. She, too, was a Comet stewardess waiting for a flight to the Far East. She suggested that we meet at Swan and Edgar's and then treat ourselves to a little shopping spree in the city to lighten my mood. I thought this was an excellent idea, a distraction that might whisk away the black cloud hanging over me. We both loved dresses and hats, the former to be chosen for warmer climates. It was a fun day scouring the shops and dressing up. I chose a smart pale-blue sleeveless summer number with a round neck and full circular skirt from Debenham's. I also bought a gorgeous red pillbox hat that looked stylish with my gray coat and winter boots in the cold misty weather.

Before returning to our lodgings, we dropped into the folksy and charmingly-named restaurant *The Brief Encounter* in Knightsbridge for a coffee and a chat. We found a snug little table close to an immense window overlooking the street. As we sipped our coffee, I glanced at Patsy, then lowered my eyes, placing my cup onto its saucer, sighing. She knew what I was thinking.

Cool and composed, Patsy, reaching over, put a hand on my shoulder and looked at me with her big, blue eyes. "Stop worrying," she said, trying to snap me out of my gloom. "And stop forecasting the worst. It could be something really trivial."

The weather on the morning of the 4th was bitter. After ensuring that my uniform looked spick and span, I grabbed my heavy company coat and waved goodbye to Mabel and Ossie. They were in suspense on my behalf, as I had told them about the meeting after receiving the ominous call. I shot down the stairs out onto the cold, wet pavements, to catch the train to the airport.

As soon as I arrived, I was immediately whisked into Mr Drayson's office. "Good morning, sir," I said breezily, trying to hide my nervousness.

Just finishing a phone call, he motioned for me to take a seat. "Good morning, Miss Baker."

I sat in front of his desk, watching his movements and penetrating eyes frisking me over. He came straight to the point. "How would you like to go to America?"

I was astounded as Comets were not flying to America at this point. Did he intend to transfer me to another fleet? I hoped not. I was happily settled in London and loved flying on the Comet.

"I would love to go, but what is the reason?"

"We want you to represent BOAC at the World Travel Show in Chicago to publicize the Comet, the jet age."

My 23-year-old eyes grew wider. I couldn't believe what I had just heard. Were my ears deceiving me? This was overwhelming. "Oh, thank you so much." I was, in fact, lost for words.

He wasted no time in relating all the information in his rapid patter, giving me a long list of things I had to do, to prepare for the trip. "We will get the American visa and any other essential documents for you. So, I will need your passport."

Fortunately, I had my passport in my handbag since I always carried it with me when in uniform. As I handed it to him over the desk, he said, "Return to my office on Friday, 6th of November for a final briefing and to collect your travel documents. You will leave on Sunday the 8th of November at 8pm on Monarch Stratocruiser, service 509/944."

It seemed unreal being tasked with such a glamorous mission. I thanked him again and virtually floated out of his office, ecstatic, the butterflies swirling around in my stomach. It would be my first time in America. I'd only ever seen pictures of New York in movies and magazines and dreamt of going someday during my flash-in-the-pan modeling stint. On the tube home, my mind flickered with images of gothic brownstones with geraniums and cats in the windows, avenues lined with flashing neon lights and lit- up storefronts, ice skaters amid the winter wonderland of the Rockefeller Center.

When I rounded the corner to Old Brompton Road, I sprinted up to the house and couldn't wait to tell Mabel and Ossie about my exciting new assignment. "What a gift," Mabel said breathily, "going to that amazing city... that island in the center of the world." Ossie put on Little Richard's new record Boogie and said, "That's marvelous news." They were thrilled, opening a bottle of wine, as the three of us enjoyed a small celebration. Unfortunately, my flatmate, Jo, was away on a flight to Basra.

In the interim, from the time I left the airport to my arrival back at the flat, the press had received news of the BOAC publicity drive. A few weeks before, Rex North, a popular *Daily Mirror* journalist, called me his 'sweetheart of the air' in his column and the paper printed a snapshot of me in uniform. I had woken him up before landing in London after his trip to Capri and Ischia. "*Ciao, carissima*," he'd said groggily. He later wrote a piece about catching the Comet back from Rome when "it screamed like something out of hell" and being roused at 4am by the stewardess "who seemed to think it was a good time to drink coffee", but found it hard to be annoyed "facing a radiant smile". I was delighted.

On Friday, the 6th of November, I spent the morning packing for my trip to the States when, unexpectedly, I got call from Rex North, the *Daily Mirror* columnist. He made a point of tracking me down and invited me out to dinner the day before my departure for America. He thought we should celebrate in honor of my mission to Chicago.

So, the following night, I met Rex at Quaglino's in Mayfair, an elegant haunt of the rich and famous since the 1920s. The sumptuous stylish design was breathtaking. The owner whisked us through the showy Gatsby-esque dining room with its sweeping staircase to a secluded corner table on the first floor. Rex, somewhere in his 40s, was dapper, witty and quick with a quip. As we sat chatting and sipping champagne before dinner, photographers appeared and snapped several pictures of us gazing at each other with cocktails in hand. Afterwards he dashed out another story, beneath the headline: *Rex North Dates his Hostess with a 4am Smile*.

Time waits for no man. Sunday soon arrived. I made sure I was at the airport several hours before departure in the event of any further last-minute instructions. While hanging around in the departure area, a photographer and reporter came up to me, asking to take my picture and interview me about my forthcoming American publicity tour. So, with time to kill, we convened in the bar area, an elegantly modernist space. I ordered a tomato juice since I was on the clock while they enjoyed martinis and endless cigarettes, as we discussed the jet age project at the Sherman Hotel in Chicago.

~

ONCE ON BOARD THE STRATOCRUISER, I settled into a comfortable window seat. No one was in the seat beside me, which meant I could stretch out when I wished to sleep. There were three stewards and one stewardess on this fight. They were a fabulous bunch who treated me like a celebrity, with no end of flowing cocktails and delectable cheeses and cookies. After the lights went out, I was full of excitement, couldn't sleep and chatted with Miss Grantham, the stewardess. Her face lit up as she raved about New York, her adopted home. She loved the anonymity of the city, its fabulous and frenetic energy, brimming with possibility. As I listened, wide-eyed, the smoky live-music cafes of Greenwich Village, the glitzy shows on Broadway and the beachfront sliver of Coney Island flashed before my eyes like scenes from a moving train. As flight crew, we had a 'globalized' existence, with exotic layovers, and shared stories about the various far-flung places we'd seen and wanted to see.

Our route took us from London to Shannon, a flight time of two hours and seven minutes. Next Shannon to Gander, where we were delayed for a few hours. A minor technical problem grounded us in Gander. Then, from Gander to New York, arriving in Manhattan during the afternoon of the 9th November. I wrote all the information in my flight logbook which I always carried with me.

As soon we landed at Idlewild Airport in New York, I found out the BOAC representative couldn't meet me due to heavy traffic delays. I was given a message from a BOAC customer-service agent asking me to take a taxi to the Shelton Hotel in town. Here, I would spend the nights of the 9th and 10th at the expense of the company.

A staff member accompanied me to the exit at the airport and ensured I was seated in the transport. The driver, named Sam, was well known to the staff. He was a delightful man, whom I discovered had a flair for humorous chatter. On our way into town, I was enthralled by everything we passed as the taxi slowly crawled along the traffic-jammed road. Suddenly, a colossal high gray wall came into view, stretching for several blocks, and mesmerized me. Behind the enclosure, the top section of an enormous cross silhouetted by the sky protruded above the wall.

"Sam, what is behind that massive rampart?"

"Ma'am, that is a c-e-m-e-t-e-r-y. They had to build a great wall around it because the people are simply dying to get in," he answered in his American drawl.

The remainder of the journey was spent telling me interesting stories and anecdotes from the years of his early life. During our brief encounter, I became quite attached to this funny, witty guy, who was not only intelligent but entertaining.

At the Shelton Lexington Avenue, Sam stopped his taxi in front of the hotel and carried my suitcase into the foyer, where Peter, the BOAC agent, was waiting for me. And during our cursory conversation, he suggested we have dinner at a nearby restaurant.

I felt a little flat and jetlagged. "That would be lovely, but I would like to get an early night as I'm tired having flown for 16 hours."

"After your long flight, you must have a bite to eat before retiring," Peter insisted. "I'll make sure you're back at the hotel within a reasonable hour."

A shower and change of clothes made me feel refreshed and gave me a sudden jolt of energy. I met Peter in the lobby downstairs. He took me to a charming, folksy little Italian restaurant hidden away in a side street with ivy growing over a trellis at the entrance to the front door. In the seclusion and over our candlelit table, we discussed my scheduled events for Chicago. A violinist passed by fiddling *Trieste Mia, Te Voio Ben* and other romantic tunes. All reminders of the voyage on the Italian vessel, HMHS Gerusalemme. The time my sister and I sailed along the east coast of Africa to Venice.

"Tomorrow, as we have the day at our disposal, I'll show you several important places in and around the city," Peter promised. "Anything you'd like to see?"

"I would love to visit the Empire State Building, Rockefeller Center and Grand Central Station," I rattled on excitedly. "Then Central Park, Fifth Avenue, the church from where the Easter Parade begins and the Waldorf Astoria. After seeing the film *Weekend at the Waldorf*, I've always longed to see the Starlight Room. It used to be the gathering place for the rich and famous and still is. Hmm... and I've been told the Art Deco of the hotel is magnificent."

He laughed at my wild enthusiasm. "Well, we will be engaged, but I doubt whether we'll be able to do all that. Anyway, let's see how many places we can manage. And now, young lady, I suggest we return to the hotel so you can have a restful sleep to restore your energy for tomorrow's excursion."

The dinner had been relaxing, interesting and very pleasant, but I was pleased to return my room. As soon as I got back, listening to the steady, soothing sounds of the buzzing street outside, I sank into bed and immediately drifted into a deep sleep.

The following morning, dressed in BOAC uniform, Peter met me in the hotel lobby. We ventured out in overcast weather. Driving through the highways and byways, my mind boggled at the immensity of the city. There was concrete everywhere you turned. Our first stop was Fifth Avenue, which started north of Washington Square and stretched all the way north to Harlem. Peter informed me that this was one of the world's most expensive streets that had everything and anything. But I let him know that I was more interested in museums and churches. We headed for the magnificent Neo-Gothic structure of St Patrick's Cathedral between 50th and 51st Streets.

I was captivated by the ornate architecture. "Peter, it's awe-inspiring." Then my thoughts side-tracked. "I can envisage Fred Astaire and Judy Garland leaving the cathedral to join the famous Easter Parade. Then stepping along Fifth Avenue, accompanied by a band playing popular melodic tunes."

"Do you know that the Easter Parade started as far back as the 1870s? And from the 1880s it was one of the main cultural expressions of Easter in the United States. It had also become a vast spectacle of fashion, apart from the observance of religion."

He held my arm and led me through the main portal into the interior of the church. I gazed at the magnificent surroundings. Then we proceeded to the Lady Chapel with stained-glass windows crafted in England.

The Pietà statue near the chapel caught my eye, so I walked towards it for a closer look. "It's larger than Michelangelo's Pietà in the Basilica in Rome," Peter said.

"Oh wow! It's so lifelike and inspirational."

It was early evening by the time we had completed part of my requested outing. Back at the hotel, we arranged to meet the following morning for our journey to La Guardia Airport to catch my flight for Chicago.

∿

ON 11TH NOVEMBER the United Airlines DC-6 Convair, UAL 505, left La Guardia, New York at 11:10am. The estimated flight time was three-and-a-half hours to O'Hare in Chicago. The crew members were captain Chrisman, F/O Hunt, E/O Davies and stewardess Miss Edwards. It was a smooth flight. Gazing out the window, I marveled at the voluptuous pillowy clouds coasting past, forming animal shapes and flowers, mermaids and pirates, devils and saints, and thought about how pilots, sailors and farmers predicted the weather by the different cloud formations. After lunch, the captain invited me to view the flight deck before landing. I was always interested in comparing the cockpits and flight panels of dissimilar aircrafts. As the first officer drew my attention to some instruments, explaining how things worked, a minor problem cropped up. Not wanting to be a hindrance, I left the flight deck and mingled with the other passengers. This was a time spent playing cards, drinking cocktails and engaging in fascinating conversations.

I stepped out of the plane to blinding camera lights. Glimpsing a group of reporters and photographers waiting on the tarmac, I was shocked when I realized they were there for me. At the bottom of the steps, a cheery, moustached gentleman in well- cut suit stepped forward and introduced himself as Wally Owen, a BOAC representative who would act as my aide-de-camp shepherding me around on various press assignments. With an enormous smile, he handed me a model of the Comet and asked me to present this to Miss Edwards, the United Airlines stewardess, specifying various features of the aircraft for the benefit of the gathered news media. Again, the cameras clicked away.

Miss Edwards from United and Miss Baker from BOAC

Once the media got their news stories, Wally picked up my bags and hustled me off to the WLS Radio Recording Studio. This was hosted by Martha Crane at 1230 Washington Street. On the way there, cruising through heavy traffic, Wally handed me the schedule of engagements covering the four days I would be in Chicago.

Publicity schedule

At the radio station, I was required to be on the air at 3pm. Miss Crane, the host, introduced me in the opening minutes and then asked a few questions about the Comet and the jet age.

When it was finally over, Wally took me to my hotel to change before heading to Patricia Stevens Modeling School, run by a former Kansas City beauty queen. I had been invited to give a talk about the Comet and BOAC to dozens of up-and-coming United Airlines stewardesses. Everyone present appeared fascinated by the Comet, particularly as it was the world's first passenger jet. At the end of the session, I was inundated with questions. It was stimulating spending the next hour sitting in the lecture room, milling with the students until Wally lured me away.

On Thursday, there was a very early start. Wally made sure I arrived at the Terrace Room of the Morrison Hotel by seven o'clock in the morning to appear on the Don McNeill Show at 7:30am. It was a long-running variety radio show, which made a short- lived transition to television known as Don McNeill's Breakfast Club. Fans travelled hundreds of miles across America to attend at his hugely popular performance.

It was fun watching the TV commercials from the corner of the recording studio. Before I appeared on camera, a Smith's Toasted Crisps ad was aired. Holding up a packet, Don McNeil commented, "If you want to remain slim and trim, tuck into a packet of Smith's Toasted Crisps."

The remark passed right over my head. I made my appearance on stage for the audience with Don McNeil, a big, high-spirited fellow. After introducing me as a Comet stewardess to America and the jet age, he asked, "How do you keep so slim and trim?"

Without thinking, I jokingly said, "I eat Smith's Toasted Crisps."

The house roared with laughter and my popularity soared, not only with viewers present but with those all over America watching the performance on TV. Hours after the show aired, I was amazed at the number of times I was stopped by people asking for my autograph. This happened when walking down Chicago streets or jumping onto buses. The chic, curve- hugging blue uniform no doubt attracted attention. After all, for a time, airline stewardesses were perceived as glamorous.

In between flitting from station to studio, Wally and I had five-and-a-half hours of downtime. My next assignment wasn't until two o'clock, so he showed me some of the sights. The city's sky-high wonders unfolded, the neo-gothic Tribune Tower and Art-Deco Palmolive Building and iconic Great White Way, Chicago's theatre district, which, apparently, Wally said was a blaze of lights at night. We ended up along Lake Shore Drive, a scenic, tree-lined stretch from downtown along the Lake Michigan shoreline, gazing at the uninterrupted view of the city's soaring skyline and scattering of white sailboats gliding along the blue swells that seemed untamed and isolated in winter. I snapped some photos, then we headed back and arrived at W. Madison Street in time for my *Daily News* interview, which was short but to the point. Afterwards I headed straight off to the hotel to rest for a couple of hours before going to the Consul General's cocktail party at 5:30pm at his home in North State.

The next morning, on Friday the 13th, my *Daily News* article came out. Wally showed me the piece with a photo of me in uniform and a slash of red lipstick, my hair swept up beneath my cap. I was portrayed as an English girl from Rhodesia, Africa, "a one-woman Chamber of Commerce for the jetliner," as I had compared the smooth and silent Comet to "a magic carpet ride" soaring across continents and oceans. The piece snappily laid bare my fascinating life as a hostess up above the clouds, amid famous authors, movie stars and moguls, like a glittery cocktail party with wings. For a short spell, I was literally the face of the Comet.

Consul General's house

Mr Alcott, Consul General, Mrs Valder, Mr Valder and Mr
Wally Owen

Our host had invited 200 guests to the function. When my Aide-de-camp arrived, he introduced most of them to us both. Everyone was intrigued with my uniform and the story of the Comet. Questions were never-ending. It was a sensational party held in a beautiful home. By the end of the night, I was feeling a little weary as I never stopped talking. When several guests were leaving, Wally approached me suggesting we follow suit. "You must be exhausted," he said.

"I am a little tired after all the non-stop excitement."

We sought our host, who was busy conversing with other guests, and thanked him for a fabulous evening before leaving. "The Consul's chauffeur is taking us back," Wally explained. "He'll stop at your hotel and then drop me off at my home."

We drove through the busy streets, and as I watched the bright lights of the traffic, turning like a snake on the road, I was feeling drowsy. When we reached the hotel, the porter opened the car door for me.

"Sleep well, you have a very active day tomorrow," said Wally, bidding me goodnight as I got out of the vehicle.

FRIDAY – what a hectic day! Checking the list on my schedule for this day:

Press Preview at the Sherman Hotel – 09:00.
Louise Wright – WMAQ Radio, Merchandise Mart. 19th Fl. (It's a Small World) – 11:00.
Hi - Ladies WGN – TV, 441 N. Michigan – 11:45.
D.B.E. Luncheon – Midland Hotel, 172 W. Adams St. (Persian Room) – 12:30.
Mrs. Charles Walgreen – 3240 on Lake Shore Drive. Cocktail Party – 15:00.
World Costume Ball, Sherman Hotel (must appear in uniform) – 21:00.

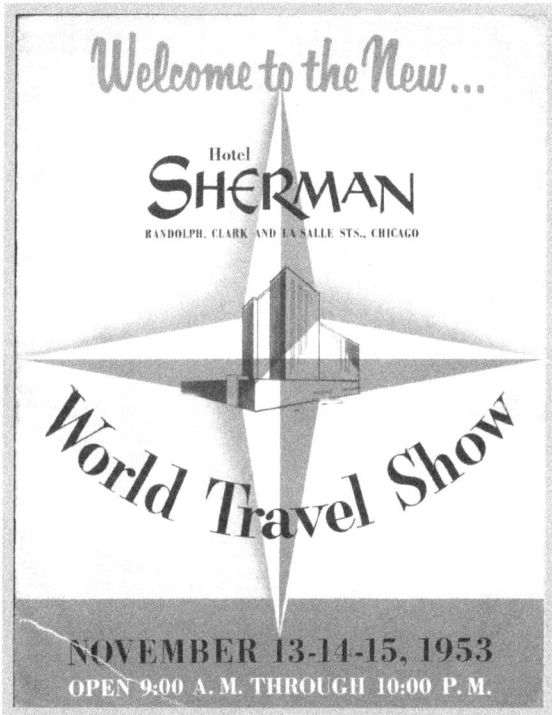

Sherman Hotel travel Show, Chicago

BETWEEN ALL MY ENGAGEMENTS, I had to make appearances for promotional press photos and interviews. These took place at the World Travel Show, BOAC booth in the Sherman Hotel. Also, I had to mingle and talk to the public about the Comet, the jet age plus answer questions.

One afternoon while standing at our stall, busy chatting to passers-by, Wally appeared along with a tall, well-built and distinguished-looking man. He had a happy, handsome bronze-skinned face. I could hardly believe my eyes. I was looking at a real Native American wearing full regalia of traditional dress. He shook my hand, beaming and showing his perfect gleaming-white teeth, and introduced himself as Chief Oshkosh of the Menominee Tribe. He spent time with me in the BOAC section while I pestered him with questions, which he cheerfully answered. Chief Oshkosh told me he had built and owned a Trading Post in Egg Harbor, Wisconsin, and was now publicizing his business at the World Travel Show. He was intriguing but soon Chief Oshkosh had to return to his booth as an important visitor wished to speak with him. Before leaving, he printed his name and address on the back of my schedule sheet — something I will always treasure.

Chief Oshkosh and Miss Baker

Wally arrived to take me to my hotel after spending several hours at the Travel Show. And having previously attended Mrs Charles Walgreen's afternoon cocktail party, it gave me time to rest and get ready for the World Costume Ball, in uniform, at 9pm, back at the Sherman Hotel. When I entered the ballroom, I stood for a few moments, gazing around at the scene before me. Two lines from a Wordsworth poem came to mind, out of context, but nonetheless... *"I gazed – and gazed – but little thought, What wealth the show to me had brought."*

The grand, ballroom had shimmering crystal chandeliers hanging from the high ceiling and was exquisitely decorated. A live orchestra played classical music, intermingled with popular wartime melodies. Attendees wandered around in a worldly array of costumes. Some held masks in front of their faces and others waved small or large brightly colorful fans. But they never ceased to stop and talk to me, dressed as I was in my BOAC uniform, about the beautiful Comet and flying in today's jet age. By the end of today, it was a pleasure to reach my hotel room where I finally fell into bed after midnight.

~

ON SATURDAY THE 14TH, my last two appointments were on the cards. As usual, Wally collected me that morning as we had to be at the Lake Forest Reception by 11am for yet another cocktail party. Photographs were taken while circulating among the guests, telling them about the comforts of flying in the Comet. Wally left me to do the rounds. I was never lost for company. People were still astounded at the fact that I was a stewardess on a passenger jetliner that traveled at 400 miles per hour at a height of 40,000 feet..

Then Wally, my aide-de-camp, interrupted at an opportune moment and suggested it was time to leave. He took me to lunch at a lively, back-street French bistro. This was our retreat, where we could pass the time for an hour, including a pleasant meal. A jovial gentleman sat at a grand piano in the corner of the room, playing sentimental French tunes, including *La Vie En Rose* and *J'attendrai*.

This whirling dream of a tour ended in this quaint little restaurant off the beaten track. After lunch we wizzed off to my final appointment at the Merchandise Mont, a half hour radio recording.

ON SUNDAY MORNING THE 15TH, back in Manhattan, I was again met by the BOAC representative Peter, who said I had a full day in the city before my late-night flight to London. He had planned a sightseeing bus tour. He parked his car at the Waldorf Astoria Hotel. From here, later in the evening, I would catch my transport to the airport for the final flight back to England.

As we waited at the hotel entrance, a small tour bus arrived. Driving through a section of the Bowery, it was disturbing seeing so many homeless people sprawled out in doorways and on the pavements. We passed cheap cafes, dance halls, saloons, flophouses and pawnshops, patronized by the poor locals and vagrants, decent, ordinary people with hardscrabble lives. Some people who walked in the area had to be mindful of gangs frequenting the neighborhood. It was a sombre scene.

We continued along the highways and byways of New York, arriving at Ellis Island, home of the Statue of Liberty. She looked so regal, standing on her towering pedestal overlooking the gateway to America. Her torch held high to the heavens.

"What does the tablet in her left hand signify?" I asked Peter.

"That is symbolic of the Declaration of Independence," he explained. "The statue, a gift from the French, was constructed in Paris and finally completed in 1884, then brought to the States in 214 crates."

"Can one walk to her crown?" The immense size of the viewing platforms were intriguing.

"If you can manipulate 354 steps. Do you feel inclined to walk all of those steps to her crown to view New York City? It's an impressive scene from the top if you're interested," Peter said, smiling.

"Thanks for the invitation, but I don't think I could manage the ascent today."

Finally, our tour ended back at the Waldorf, where I waited for the bus to take me to the airport for my overnight flight to London.

Peter and I had become good companions and were sorry to bid one another farewell. It was sadly part of the job, forming instant relationships with people we'll likely never see again.

"Ships that pass in the night, we might meet again someday," I said to my friend.

"Keep in touch between busy moments, if possible," he replied, giving me a hug.

On a chilly night, we took off from Idlewild Airport (JFK)

just before midnight. I slipped into my seat and immediately fell asleep. I was back in Sri Lanka with Brian. I dreamt we were alone dancing in a courtyard with paper lanterns glowing red and orange off dark, shadowy trees beneath a full moon. When the stewardess woke me at 6:30am in the morning just before landing in Newfoundland, the pilot announced that there was a mechanical fault with the aircraft. Flying low over miles and miles of rocky pine-cloaked wilderness, with a few brightly-colored wooden houses and weathered fishing boats clinging to the wild, windswept shores of the North Atlantic, I thought of all the amazing wild things that called this home — the moose, lynx, arctic foxes, puffins, caribou — and wondered if that meant we would be stranded here for a while.

The atmosphere was heavy in Gander. This rugged Canadian small town was an international refueling station and would become known as the crossroads of the world, where we had an overnight diversion. Our aircraft finally took off the following afternoon and we arrived in Shannon in the evening. Once again, we were told we were spending the night due to a further technical hitch. I was wary, at first, of what this hitch really meant. Luckily I wasn't in uniform and indulged in a few cocktails and tried my first Irish Coffee. We all had a few whiskies and a good old Irish sing-song before dinner.

Midday, we took off from Shannon in misty, miserable weather. The patchy fog subsumed everything. Again we were forced to land in Bournemouth as the visibility at London airport was poor. Everyone accepted the situation and seemed to revel in the fun, friendship and fleeting little vacation. I was happy to be back in London and looking forward to doing nothing. But when I got back to the flat, Ossie, Mable and Jo threw a surprise welcome home party for me and had invited about 30 of my co-workers and friends. It was the perfect ending to a fantastic trip.

COPY

British Consulate General,
720 North Michigan Avenue,
Chicago 11, Illinois.

November 18th, 1953.

Dear Leslie,

I must congratulate you on the success of the BOAC participation in the World Travel Show at the Sherman Hotel from November 13 - 15. The presence of your charming Comet Stewardess, Miss Elaine Baker, was a brilliant asset.

Mr. Rodney Chalk, of BIS, who handled much of the British Travel Show promotion, tells me that she was an outstanding success on the various radio and television shows and press interviews arranged for her. From my own observations when I visited the show on Friday, I would say that she was not only doing a splendid job for BOAC, but also for Anglo-American relations.

Sincerely yours,

Berkeley Gage.

Leslie Valder, Esq.,
District Sales Manager,
British Overseas Airways Corporation
37, South Wabash Avenue,
Chicago 3, Illinois.

Letter from the Consul General 18 November 1953

Letter from District Sales Manager, Chicago 20th November 1953

BRITISH OVERSEAS AIRWAYS CORPORATION

STRATTON HOUSE, PICCADILLY, LONDON, W.1

Phone: MAYfair 6611 Telex: VICtoria 3126 Telegrams: Speedbird Wire London

B·O·A·C

HFP.3396

27th November, 1953.

Dear Miss Baker,

    I am sure you will be interested to know that our Public Relations Officer in North America, Mr. Wynne, has written to say that everyone in New York and Chicago were delighted with what you were able to do on behalf of B.O.A.C. on your recent visit to the States.

    Our District Sales Manager in Chicago has written to say that they were afraid they kept you extremely busy but that you proved to be "a most adept ambassadress, whose appearance, charm and self-possessed manner completely captivated everybody."

    May I add my personal thanks for the very splendid performance which resulted in wide and excellent publicity for the Corporation in America.

              Yours sincerely,

              F.A.Allcott
        D/Chief Press and Information Officer.

Stewardess Elaine Baker,
c/o the Comet Fleet,
London Airport.

Letter from Chief Press and Information Officer 27th November 1953

BRITISH OVERSEAS AIRWAYS CORPORATION

37 SOUTH WABASH AVENUE
CHICAGO 3, ILLINOIS
TELEPHONE FRANKLIN 2-3987

HEAD OFFICE
LONDON

AND IN
BOSTON, DETROIT, LOS ANGELES,
MIAMI, SAN FRANCISCO, WASHINGTON,
MONTREAL, TORONTO

EXECUTIVE OFFICES
342 MADISON AVENUE
NEW YORK 17, N. Y

REF:CGO:CS:392b   December 14th, 1953.

Miss Elaine Baker,
165 Old Brompton Road,
London, S.W.5.

My dear Elaine,

        You no doubt wonder what has happened to all the
material which I promised you.   I am enclosing some of it now
with the promise that there is more to come.

        First of all we have not as yet received the newspaper
cuts from papers outside Chicago but they have been requested and
we shall forward them to you as soon as they are received. Secondly
your final broadcast will be heard on the National Broadcasting
Company December 27th and they are making the transcription of
this which I shall forward to you when we receive it.

        I heard the first one we did with Martha Crane and I
must say that it did come over extremely well.   The little problem which you and I had regarding the time from London to Tokyo
was not even noticed and as we talked about slipping immediately
following the mentioning of the time, it gave the impression that
that was the time which it took you to make the trip and not the
aircraft.

        The promotions were eminently successful and the results
have been most gratifying.   It might interest you to know that
our Round-the-World business has increased at an alarming rate
and I cannot help but feel that you were the primary force in this
increase.   My only wish is that we could have you make a monthly
appearance but then if we did that, I am afraid we would have to
get increased staff to cope with the increased business and my
already numerous grey hairs would become more evident.

Letter from Wally 14th December 1953

BRITISH OVERSEAS AIRWAYS CORPORATION

37 SOUTH WABASH AVENUE
CHICAGO 3, ILLINOIS
TELEPHONE FRANKLIN 2-3887

AND IN
BOSTON, DETROIT, LOS ANGELES,
MIAMI, SAN FRANCISCO, WASHINGTON,
MONTREAL, TORONTO

HEAD OFFICE
LONDON

EXECUTIVE OFFICES
342 MADISON AVENUE
NEW YORK 17. N. Y

Miss Elaine Baker   -2-   December 14th, 1953.

As a matter of interest, Ed McKean called me to inquire whether you were still in town as he wanted you back on the McNeill Show due to the fact that they had received some telegrammes with proposals of marriage. I feel certain that if at some future time you have the occasion to come to the U.S.A. you will probably find husbands waiting on bended knee.

I cannot thank you enough for the superb effort you made and I do hope that you did not go back completely exhausted.

As I have said, I shall send you the recording in due course and I trust that I shall have the opportunity of seeing you when next in London.

With warmest regards,

Wally

Letter from Wally 14th December 1953

FINAL FLIGHT G-ALYP

O n a frosty December morning, I didn't feel like
getting out of my warm bed. But the thought of
flying to the Far East and spending Christmas in the
warmth and sunshine of Karachi made me feel excited about my
departure to Singapore. I received the roster after arriving back
from my last flight on the 7th of December, and had two weeks
of leisure time in which to prepare for the trip.

The morning was zipping by. I dressed and dragged my heavy
suitcase down the stairs, having packed it the previous evening. I
would be away for two weeks, depending on whether we
encountered any minor problems or violent weather en route.
Therefore, a variety of garments filled my case, particularly as
the temperature in Rome would be as cold as the UK and we will
be 'slipping' in the city for 24 hours.

Mabel and Ossie were in the ground-floor studio, busily
working in their darkroom. Leaving the flat, I called a hasty
goodbye and wished them Happy Holidays before shutting the
front door. They were accustomed to Jo and I flying in and flying
out, so formalities were always disregarded.

At the airport, I headed for operations to check on the
passenger list and find out if there were any special requirements

for the flight. I collected my cosmetic bag containing creams, moisturizers and atomizers for the ladies' and gents' toilets. My accompanying steward Frank Saunders arrived soon after me. He went through his necessary checks, including signing for the portable bar and its contents. It was most important that a meticulous count of all drinks was listed on every flight. Frank and I strolled across the tarmac to our aircraft G-ALYX to wait for the flight-deck crew and passengers to board. Every time I looked at the Comet, it sent chills down my spine. Even after all the months having flown in her, I still thought it was the most beautiful aircraft I had ever set eyes on. She was the ultimate airliner.

"Oh, Frank! Look at her, isn't she gorgeous standing in the snow with her perfect lines? She is so alive."

"Come, girl, pull yourself together," he said, laughing. "It's just another aeroplane."

We climbed the stairs leading to the vestibule. After hanging up our heavy coats, I went about my work setting up the ladies' and gents' toilets with the Elizabeth Arden accessories. Frank headed for the galley to prepare the tea trays for after take-off service.

There were no VIPs to impress on this flight, although everyone was unique and treated alike. As soon as the passengers were seated and the flight-deck crew had boarded, I walked around the cabin handing around cotton wool and barley sugar. We departed at just after two o'clock for Rome, soaring through heavy gray clouds into the vast blue sky.

When the correct altitude had been reached, the seatbelt and No Smoking signs were turned off. The passengers submitted to the casual atmosphere of the cabin. Some writing postcards or playing chess, others removing a cigarette from their silver or gold cigarette cases. Although the stewardess distributed cigarettes to them as part of the service. Thirty minutes after take-off, we served tea or coffee, sandwiches and cake. If a passenger requested something a little stronger, Frank opened the bar and produced whatever was desired. After serving the passengers, I returned to the galley and took refreshments to the flight-deck crew. My captain was Alan Gibson, who won a DFC (Distinguished Flying Cross) during the war and had flown 4,267 hours with BOAC. I had met him previously when he was a supernumerary captain on a flight to Singapore and before that, when he was my captain on a London-to-Colombo flight. The rest of the crew were first officer William John Bury, engineer officer Francis Charles McDonald and our radio officer Luke Patrick McMahon.

Sitting in the stewardess' seat, before descent, I spent a few minutes filling in my logbook and the handing-over report for the stewardess taking over from me. As the Comet touched down, I stood at the front of the cabin and gave the usual transit briefing. Some would connect to a flight for other destinations.

Once the plane came to a standstill, the passengers left the aircraft. I accompanied them as they strolled across to the terminal building. It was important to ensure those in transit were seated in the restaurant before I returned to G-ALYX to complete handover procedures. After chatting with the steward and stewardess taking over the aircraft for the next leg of the flight, Frank and I made our way to the main building, collected our per diems from the BOAC miniature bank at the airport and caught the company bus that would drop us all off at our respective hotels in the city.

It was interesting to, again, be in this historical environment for 24 hours. So much to see in the short time available. On the bus, everyone seemed deep in thought. Mr Bury, the first officer, broke the ice and asked, "Anyone care to join me for a drink and dinner in the Quirinale pub later this evening?"

"Will meet you later," said Frank, replying to the invitation. He was in a happy frame of mind.

I was tired when we reached our hotel. So, once in my room, I bathed, changed into my pyjamas and climbed into crisp, white sheets. Then, picking up the phone, I ordered my dinner to be sent to the room.

The following morning was spent at one of my favorite stomping grounds, the Spanish Steps. I walked across the Piazza di Spagna to the monumental stairway, which was created by the architect Francesco de Sanctis in the 1700s. Looking up at the church at the top, the Trinita dei Monti, I realized it would take a few hours to view its interior. So I decided instead to meander along to the Trevi Fountain, where I returned on every stopover in Rome. Each time, never failing to throw three coins over my left shoulder, into its waters. Reflecting on the time Brian and I shared romantic moments throwing coins over our shoulders, hoping to meet here again. The sound and colorful beauty of the Trevi waters always left me lost in thought and made me think of Brian.

Turning away, I headed toward the river and Castel Sant'Angelo. It was an enjoyable walk passing a variety of scenes en route. After all, this was Rome. Castel Sant'Angelo was built as a tomb for Emperor Hadrian and is known as Hadrian's Tomb. Later it was a burial place for Roman emperors, then a fortress for the popes. In the world of today, it was a museum, where I would spend a couple of hours brushing up my Roman history.

～

LEAVING Rome on the 23rd of December, we flew via Beirut and Bahrain to Karachi. On take-off, Frank was strapped into the jump seat at the front of the aircraft, while I took up my position in the rear of the cabin. We ascended at a rather steep angle. Looking up at the steward's panel, I noticed his light flashing erratically on the board. Something was wrong. Once the seatbelt lights were extinguished, I made my way to the galley and reaching the closed door, observed a few peas rolling beneath it. Pushing it ajar, I returned the peas with my foot, hoping that none of the eight passengers in the forward cabin had noticed. I then poked my head around the door and found Frank, looking a little frazzled, in a total dilemma. Disaster. One of the containers holding the sky plates had not been secured. On take-off, the container door had come away, scattering all the dinners over the floor. The passengers could not go without dinner.

"Darlin' please take care of the passengers while I salvage this mess," he said, "Dinner will be late. Offer them extra drinks on the house but keep them entertained. I'll let you know when the service will start. In the meantime, I'm preparing cocktails and canapés which I would like you to dispense. The snacks should keep them happy for a while."

"What are you going to do? We don't have extra meals."

"I will have to retrieve every morsel and return it to the sky plates, after all the floor is clean. Will let you know when service is ready. Remember, give the passengers another round of free drinks."

The travelers were a pleasant group. I socialized when possible, talking to those that were in the mood for conversation. Nicholas Monsarrat, the author, was seated midship. I had recently read his novel *The Cruel Sea*, an absorbing historical story about the Battle of the Atlantic during World War II. I was delighted to chat with the gentleman during my rounds. He appeared to be a polished, unassuming character who spent most of his time gazing out the window at the wild blue yonder. Possibly deep in thought, or contemplating his next book.

As the passengers finished their drinks, I offered a further free round. One well-dressed gentleman requested a whiskey and soda. So, I opened a miniature for him and poured the whiskey into his glass. Then went to the galley to retrieve the large crystal soda syphon, which was rather heavy. Pressing the syphon handle, the soda shot into his glass, causing a massive surge rather like a fountain. Spraying the whiskey and soda all over his jacket, shirt and tie. This had never happened before, but the handle was stiff and needed extra pressure.

"Oh, sir, I'm so, so sorry." I was mortified. "Please have everything cleaned and send the bill to BOAC."

This passenger was courteous and laughed, finding the incident amusing. "Don't worry, it will soon dry." He couldn't have been kinder.

Soon after the incident, Frank and I served dinner. All the passengers remarked on the delicious meal. After the coffee and liqueur service, we dimmed the cabin lights so they could succumb to the peace of the environment.

Our aircraft finally arrived in Karachi at three o'clock in the morning on Christmas Eve. When we, the crew, reached Speedbird House, I went to my room and, as usual after a night flight, fell into bed.

BOAC rest-house in Karachi

The following morning, while in a deep sleep, the ringing of my alarm woke me. Still half-asleep, I stretched out my arm and switched it off. I got out of bed drowsily, had a steaming-hot shower, dressed and hurried into the dining room, where I joined captain Gibson for a late breakfast. In conversation, he asked, "Would you care to go with me into town? I want to buy a few things for my family for Christmas, though we'll only arrive home after the festivities."

"I'd love to go along. I want to buy a sari to wear for our Christmas party tonight."

This wonderfully warm, sunny day made one feel energetic, so I was happy to explore the colorful shops and bazaars. Once we reached the shopping precincts, we parted ways, meeting a little later. Meandering through the center, I found a glamorous peach sari in a soft georgette material, decorated with sequins and gold beads. I tried it on in the shop, so that the attendant could show me how to pleat the waist and adjust it around my body. The sari toned well with my natural dark-blonde hair, and its ethereal softness looked quite stunning. Then captain Gibson appeared to let me know the transport was waiting to take us back to the rest-house.

"I have found nothing interesting but will try again in Bangkok. Your sari looks fabulous," he said on the way out.

"I love the color and am quite delighted with it."

We hurried to the waiting transport chatting about all the unusual bargains one can buy in the Karachi markets.

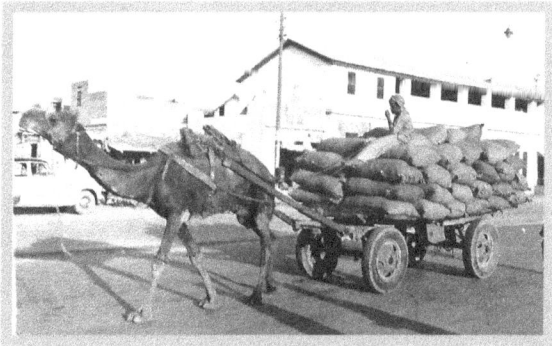

Karachi transport

Mr Bury, our first officer, had reserved a table for us in the evening of the 24th, at the renowned Beach Hotel for the Christmas party. Another Comet crew returning to London were joining us, and my friend Patsy was the stewardess of that flight.

It was a balmy evening with a large moon and bright stars in the clear tropical sky. Tables were set on the terrace with an Indian band playing soft background music that added to the mellow atmosphere. Our group comprised six males and two females, the other two crew members were visiting old friends in Karachi.

Christmas dinner was cooked to perfection and well presented. We finished our meal with the usual Christmas pudding, accompanied by a full-bodied brandy sauce. I sat toying with the pudding. Then Mr McDonald, our engineer, got up from the table and said, "Well, girls, you have no shortage of dance partners, so let's trip the light fantastic." He took my hand and led me onto the dance floor.

We all spent a fantastic evening together and talked and danced into the small hours of Christmas Day. The crew members regretted not being with their families at this time of year. Frank missed his wife and daughter and the excitement of Rudolph, with Santa Claus, dropping off presents.

❧

I SLEPT in until late morning on Christmas Day, then around noon heard a tap on my door. The radio officer poked his head around and said, "Good morning, just to let you know we're all going to the swimming pool to cool off. Would you care to join us? The bus will be here in 45 minutes."

It was a sweltering-hot day and frolicking in the cold water was sheer joy. The swimming pool did not look hygienic, as it was a greenish color with snatches of algae floating about. Every moment was enjoyed splashing, playing games, jumping in and out of its depths. Lying around the surround sunbathing, and then diving again into the water, we collected the coins that had been thrown into the water. We finally got back to the rest-house late in the afternoon, exhausted from the tropical heat and humidity and reveling in the pool. Everyone was tired, so they went about their own business.

That night, I didn't feel very well. I was hot, cold, and kept breaking out in intermittent sweats, accompanied by a severe headache with slight pain in my ears. I thought I might have a touch of the sun, so I swallowed an aspirin, hoping it might be an instant cure. Because of lack of air conditioning, I started the ceiling fan and stretched out on my bed with a book but drifted into fitful bouts of sleep.

At 9am on the 26th, the phone on a small table beside my bed rang, rousing me from a sound sleep. I lifted the receiver. It was Carl Pitu, the first officer I met on my inauguration flight. He and his crew were on their way back to the UK and had a two-day break in Karachi.

"Good morning, Elaine. I realized you were here from information in the aircraft logbook." He had an appealing Canadian accent. "Anyway, several of us are sailing across to a small island off Karachi this morning. Myrna, our stewardess who you know well, wondered if you would like to join us."

"It sounds like fun, but at the moment I'm having problems with my ears and thought I would remain in bed today."

"A different scene might make you feel better," he said, trying to persuade me. "Do come along. We'll meet you outside the reception office in 30 minutes."

After careful thought, I dragged myself out of bed and joined my friends for the day, as it might take my mind off my current predicament. I was pleased to see Myrna, as our paths hadn't crossed since the training days. Apart from Carl, she introduced me to the rest of her companions, engineer Naylor, radio officer Corlett and her steward Micky Birch. Her captain was spending his time with old friends who owned a colonial-style house close to the beach. A small crew bus picked us up and drove us to the Karachi waterfront. Here we climbed into a Bunda boat, a large sailing vessel with two masts which can carry at least ten passengers. It was operated by several local sailors who, are experts at their work, so one feels quite safe in their hands. To equalize the boat, a plank is balanced out over one side, riding above the sea. Then one or more sailors rush onto the plank to weight it down, and one of them remains sitting on it to equate the balance. Primitive, but it works.

A bunda boat

Man stabilising the bunda boat

The sea, which was green and wild, became choppy and the short crossing took about an hour. As we reached the island shore, the sailors leaped off the boat and then, with long, thick ropes, dragged the barque partly onto the beach. It was stabilized by pushing substantial wooden stakes into the sand, around which they tied the rope. We jumped into the knee-deep water and waded onto the shore.

I was intrigued by the scene that confronted me. This part of the island lacked vegetation and appeared uninhabited. Apart from tumbledown shacks and several anglers with their tired-looking camels as companions.

Main transport

It was these forlorn animals we borrowed to race along the sands. Though they were not comfortable, it was lots of fun competing and galloping through the water lapping onto the beach. Carl excelled and beat us on his ship of the desert, so we rewarded him with exuberant cheers. Myrna persuaded the camel owner to allow us to include his desolate beast in a couple of photos. The man was quite delighted, particularly as we paid him several rupees for the convenience. His dromedary looked down on us with a supercilious air. Then, after a certain amount of coaxing from its owner, the camel kneeled down onto the sand, enabling Myrna to perch gracefully on his back. I gazed at the camel's face and realized it portrayed a variety of unique, fascinating facial expressions. His large soft liquid eyes, amplified by long lashes, appeared so very distant and sad, and made me think the orbs must hold many a hidden story.

Myrna, Micky, Carl and Ed

Later in the afternoon, when we returned to our craft and cast off, the sea had become choppy, making the boat unstable on the rough ocean. This made me feel somewhat queasy as I had never been a good sailor. Ed, the engineer, who had been feeling a little under the weather all afternoon, rushed to the rear of the boat. He flung the top half of his body over the low wooden rail. Ed couldn't stomach the lurching swells and was violently ill. Then we heard a shrill cry, and all dashed to his rescue and to find out what was wrong. He looked up at us with an anguished, ashen face and said, "I've lost my top set."

Everyone leaned over the rail, peering into the depths, but the teeth were nowhere to be seen. They had sunk quicker than we could exclaim 'Jack Robinson'. Myrna looked at me with a glint of humor in her eyes, wanting to laugh at the unfortunate situation, although we sympathized with our distressed companion.

In a convincing tone, he turned to us and said, "Tomorrow, I will visit the local dentist and by means fair or foul, I will get a temporary set until I reach home."

~

WHEN I WOKE up on the 27th, the day of our departure, my ear ache had intensified. I ignored the pain and gathered together my belongings, hoping that any unwell feeling would soon dissipate. We arrived at the airport, and G-ALYP was circling the skies, preparing to land. We were flying from Karachi to Bangkok via Delhi, Calcutta and Rangoon.

As soon as the Comet came to a standstill in Karachi, I strode across the sand to speak to the stewardess before she disembarked from the plane. After our usual greetings, she gave me her handover report and informed me of one or two very minor problems that had been rectified. She then mentioned, "There's a male passenger on board who is a celiac and diabetic. As you know, he must have special meals. He also has to inject himself for diabetes." It was a proviso in the airline we always had to be aware of passengers requiring special attention.

Once crew formalities were complete and the passengers settled on board, G-ALYP soared into the air at 7:58am on the 27th of December. The length of the flight between Karachi and Delhi was one hour and 30 minutes. During this sector, we served breakfast to our 36 hungry passengers. The service ran smoothly and every morsel was consumed.

In the vicinity of Delhi, the Comet flew into the circuit for a final approach. It was a perfect landing and our captain stopped close to the terminal building. Several passengers disembarked. Those re-boarding had a brief transit break in the airport lounge while G-ALYP was being refueled. Our departure time from Delhi was 10:43am. There were still three more sectors before reaching our ultimate destination.

When we finally arrived in Bangkok, I felt a little jaded. Partly because of the pressurization in flight, my head was throbbing with pain and I felt as though it might explode. Still, as it was the festive season, the crew insisted I attend the evening party, even for a short while. Once we arrived at our hotel, ensconced in my room, I relaxed in a hot bath and rested before putting on an elegant dress for the party. Another Comet crew was at the hotel, so we joined forces for the festivity. It was fun dancing, laughing, reminiscing and talking about shopping expeditions. Apart from experiences when visiting Temple Dawn.

"We will have more time to stroll around the shops on our return from Singapore," said Jim, the first officer.

"Definitely! But I think I'll spend time in bed tomorrow morning after this party," replied Mac, our engineer.

Looking around the decorated terrace, it was impressive to see the tropical 'Santa-like' setting. A large Christmas tree, decked with shining baubles and a glittering star on top, took pride of place in the center of the dancefloor. Placed at intervals, palm trees surrounded the expanse for dancing. And colored lights framed the entire periphery. I felt as though I had walked into the land of make-believe.

As the night wore on, the pain in my head did not abate. I caught captain Gibson's attention and told him I had to retire. After saying goodnight to everyone, I ambled to my room with the strains of *White Christmas* fading into the background, a magical song popularized by Bing Crosby. One that paints a picture of holiday nostalgia, recalling many places in time.

❧

ON THE 29TH of December just after midnight, we took off for Singapore in G-ALYY, landing at Changi Airport two-and-half hours later. Here we had a quick turnaround and a brief respite before the arrival of the passengers. As the morning sun dawned, I stood outside the small terminal building, gazing at the Comet standing on the airstrip. She looked magnificent with the sun gleaming on her fuselage. Frank joined me.

"Frank, I was thinking back in time to events that took place here during the war, and stories I have read about Changi. The Japanese, who were brutal, used our prisoners to construct this airstrip. The men were forced to work ten to 12 hours a day in the blazing sun and humidity. Most suffered from malnutrition and all kinds of tropical diseases. You can almost visualize their ghosts digging the ground with pickaxes and hammering the steel girders. It's uncanny and horrific."

"You're right," he answered, "Then, when they were finally freed from Changi, there were only a few to liberate. Many died while working and some were sent to other countries as labor slaves. The POWs were also forced to build the infamous Siam-Burma Death Railway. But that's another story."

"Do you know that Ronald Searle, the satirical artist and the creator of St Trinian's School, was a prisoner in Changi? It was here that he made a series of drawings illustrating prisoners dying of cholera, and the inhuman camp conditions. And he also worked in the Kwai jungle on the Siam-Burma Death Railway."

Glancing towards the terminal building, I turned to Frank and said, "Oh, here come the passengers. Let's get back to the present."

The Comet left Changi, Singapore at 5am on the 29th of December. Then, after an appetizing breakfast service, I chatted with several of the passengers. They were elated to be flying in this wonderful smooth, noiseless aeroplane, taking them to their destinations in record time. Those in the forward cabin were enjoying balancing pennies on end, on the table, showing the smoothness of the Comet in flight. They enjoyed the fact that there was little or no disturbing noise or vibration at this altitude from our four De Havilland Ghost engines.

On my rounds, I offered duty-free items such as cigarettes, miniatures and whiskey. Also our beautiful Comet scarves with a picture of the aircraft printed on the soft, sheer voile. Free BOAC hand fans were distributed to help waft away the heat.

Complementary BOAC fan

When we landed back in Bangkok on the morning of 29th of December, the crew were thrilled to have two full days at their disposal to explore the city and wander through the markets and shops. As soon as we reached our hotel, I went straight to bed to rest my aching head and painful ears. During the afternoon, captain Gibson phoned to ask how I felt, and to let me know that he had arranged for the hotel doctor to visit me. The Thai doctor arrived early evening. After a thorough examination, he informed me that my ears, throat, nose and sinuses were severely infected, and that's the reason I was in so much pain. He presented me with tablets to be taken at specified intervals. I consumed two with a glass of water as the doctor stood beside my bed. After he left, I curled up in my sheets and fell asleep.

The following morning, at about nine o'clock, there was a resounding rap on my door. I thought I might be dreaming. "It's only me, Jim. How are you today? We're on our way to the market and wondered if you might be well enough to come along."

"Thanks Jim, but I'm remaining in bed with my head under the pillows. I'm trust that the warmth might alleviate the pain in my head and ears, and hoping the meds will take effect. Enjoy yourselves."

"We'll see you later."

The crew arrived back as evening shadows were filling my room and popped in to find out if there was any improvement to my indisposition. They were laden with parcels and so kindly gave me with a box of chocolates for a speedy recovery. I glanced at their arms filled with packages and said, "Oh, my goodness. What have you bought?"

"We'll show you when you're feeling better, on our return to Karachi, where there will be lots of time. But now we must find room in our cases for all these goods," said captain Gibson laughing, as they left my room.

∼

WHEN WE FINALLY REACHED KARACHI ON 1st January, 1954, my ear infection had worsened, although I took the pills that I had been given in Bangkok. The acute pain spread across and down both sides of my head, face and neck. But, as it was New Year's Day, a party was arranged for the evening at our favorite stomping grounds, the Beach Hotel. Although I felt like death, I would try to drag myself downstairs since the Comet crew arriving from Colombo would join us. My friend Jean Clarke was the stewardess on the flight.

As usual, it was a gloriously warm evening with clear skies and a spill of stars, shining like diamonds, scattered in the dark universe. The tables looked inviting, adorned in white and silver, shimmering under the sparkling lights. We all gathered together at a large table. Jean and I had a great deal to tell one another. We were deep in conversation when captain MacIntosh, Jean's captain, clasped her hand, leading her onto the dancefloor. For a few moments, the fantastic dinner, the conversation and the company took my mind off my immediate pain.

Once the band stopped playing just after midnight, everyone

returned in the BOAC transport to Speedbird House. Captain Gibson walked with me to my room and informed me he and the first officer were taking me to the BOAC doctor the following morning. I had no choice in the matter.

On the 2nd of January, I found myself in the GP's surgery, anxious, between my two guardians. Several other patients were looking worse for wear, waiting to be seen before my turn arrived. Attracting my commander's attention by tapping him on the shoulder and staring into his eyes, I said, "Captain Gibson, I have a needle phobia. The medic might decide I need an injection."

"Whatever he does will be for the best and will only help you, so don't worry."

When the last patient departed, the doctor led me into his consulting room and closed the door. After a thorough head-and-neck examination, he said he would have to give me a jab of antibiotics, and I couldn't fly for a few days. Producing an enormous syringe – well, I thought it enormous – he inserted the shot into one of my muscles. At the end of the ordeal, he handed me a small bottle of pills, then told captain Gibson I would be out of commission for a couple of days.

On our return to the guesthouse, captain Gibson contacted London Airport Operations, informing them I was unable to fly and had been grounded by the doctors for a short time. He had to return to Singapore instead of going directly back to London. So London operations informed him that the stewardess on the Colombo crew must change places with me. I would return to London as a working crew member with captain MacIntosh, Jean's commander.

Jean and I were both reluctant to make the change, but had to follow orders from London flight operations. As mentioned before, a powerful bond develops between crew members. Winging it around the world together, we shared hotels, meals and adventures on lengthy trips, hopes and fears, happiness and heartaches. Particularly when you have been working together intensively for two to three weeks.

I couldn't help but feel guilty for messing up Jean's plans. "Jean, I'm so sorry you have to continue the journey back to Singapore before returning home, but thanks for your help in giving me time to recuperate."

"It's a pleasure, Elaine, and when I'm home, you must come and meet my parents and spend a few days with us before our next outbound trip."

"Gosh! I look forward to it. Anyway, you are working with a wonderful crew so enjoy the flight to the Far East. I'll see you in England."

"When I arrive home, my fiancé should be back from Malta. He's an officer in the Navy," she said proudly.

I saw my crew before they left Karachi and thanked them all for their kindness and consideration with regard to my illness and wished them safe flying and sound landings.

On the evening of January 5th, captain MacIntosh and the crew left Karachi for Rome, via Bahrain and Beirut. We were, in fact, delayed leaving Karachi due to the late arrival of the connecting Comet. Strong winds were encountered while flying throughout the sector between Karachi and Bahrain, which made schedule-keeping difficult. Then, a short distance from Beirut, another adverse incident occurred. An aircraft crashed on take-off and blocked the main runway, so we were advised to use an alternative runway before finally landing in Beirut. While G-ALYU was being refueled, Steward Holling and I accompanied the commuters to the restaurant for breakfast on the ground. The plane required a high-fuel uplift, because of the strong winds we had run into during the flight. Captain MacIntosh eventually left Beirut for Rome nearly five hours late. There were several different reasons for this delayed departure. After the many mishaps, we landed in Rome, the evening of the 5th. Another 24 hours would be spent in this magical city.

Finally, on the 7th of January, I returned to England with

captain MacIntosh and his crew in miserable weather. Still, it was pleasing to reach my homely flat, welcomed by the warmth of Mabel and Ossie.

"Well, Elaine, how did you enjoy Christmas and New Year en route?" Mabel asked the moment I walked through the door.

"We all had a great time. Unfortunately, we were delayed in Karachi due to some technical hitch with one aircraft somewhere along the route. So captain Gibson had to return to Singapore. I also fell ill and, because of my ear problems, Jean, the stewardess who arrived from Colombo, joined my crew. And operations messaged captain Gibson saying I had to return with Jean's crew. My head is almost clear now after the course of antibiotics."

"By the way, your reporter friend Ray Watts from London Airport, has been trying to get in touch with you. You should ring him."

"I don't know his number, but I'm sure he'll call again."

The next couple of days, I spent unpacking and sorting clothes to take to the launderette down the road. On Sunday, the 10th of January, as the morning light appeared, I got up and dressed, walked to the Brompton Oratory, the Catholic Church in Chelsea, for early mass service. It was a blustery, wintry day, so I wrapped up, very much missing the heat of the Eastern countries. Jo wasn't expected home for another week, or she would have accompanied me to mass. With its shining candelabra, the oratory's interior looked majestic, aglow and warm. After the service, I knelt in front of a small altar, lighting two candles. One for my uncle Havelock Heslop and the other for my grandfather Thomas Heslop, both who died in World War I. Havelock was only 16 years old when he joined the forces. After his machine-gun training, he was sent to Egypt, Salonika, Mesopotamia and several other places. He never saw his family again, because during those four years the army failed to give him a furlough. My grandmother still mourned his death. I look at a photograph I have of his grave. He was buried in Basra.

However, his name and his father's name are inscribed on a large marble memorial plaque installed on the wall of All Souls Church in Bolton. It is there that their memory lives on.

After mass and a leisurely Sunday lunch with Mabel and Ossie, I fell onto my ruffled bed, burrowed deep under the blankets, and continued reading Charles Dickens' *Bleak House*. Unexpectedly, the phone resounded throughout the flat. Ossie answered it, then called up to me, "Elaine, Ray Watts is on the phone for you."

"Thanks, Ossie, I'll be down in a minute." I shouted, closing the book and darted downstairs to take the call.

"Elaine," Ray said in a clipped, subdued tone, not in his usual cheerful manner. "Have you heard? A Comet crashed into the sea off Elba this morning. There are no survivors."

I was stunned and silent, but gasped, "Which Comet? Who were the crew?"

"Captain Alan Gibson was the pilot."

In that instant, I shuddered, hot tears flowing down my cheeks. All I could think of was the tension and terror of my colleagues, and especially Jean who had taken my place, as the Comet plunged into the sea. I was devastated. Mabel and Ossie made me some tea and did their best to comfort me. Once I was alone in my room, I sobbed uncontrollably. Later that day I returned to the Brompton Oratory, prayed and lit candles for Jean, captain Gibson, my co-workers and the passengers who had lost their lives so violently. As well as the profound sorrow, I thought about this awful suddenness of fate. Joy. Love. Loss. Luck. Misfortune. How our fates whirled like a roulette wheel, the Roman goddess Fortuna snapped her fingers, and everything changed in a blink.

EPILOGUE

ELBA, 2009, THE CEMETERY

I felt a tap on my shoulder. The old caretaker was beside me and had broken my reverie. I had been sitting here for several hours, reminiscing, and dreaming about a past life. He pointed to his well-worn watch with a battered old leather strap attached to let me know the cemetery was closing and it was time to leave. I got up from the bench, and once more looked around the entire memorial, then placed the red rose I was holding on the marble slab in front of the urn. Turning around and walking away, I strolled back along the stony route lined with greenery and trees, past the shrines and graves.

Although still feeling dazed from the heat of the sun, I turned and cast my eyes over this vast centuries-old burial ground, and wondered if any spirits could be drifting along the pathways. The stillness was eerie. The whine of jet engines seemed to penetrate the air. My attention was drawn in the memorial's direction, where high in the sky, I thought I saw a flash of the Comet hover above. Glancing skyward, in a split-second, the image vanished from sight. Was it a figment of my imagination?

Approaching the exit, the dear old custodian of the crypts and graves reappeared. Looking at me smiling, he half saluted and closed the gates behind me. It was another place, another time.

Visiting the memorial was a moving, soul-stirring experience. Being in this lonely, beautiful place gave me time and space to reflect and to sift back through my recollections, to think about the many inspiring people I met and worked with and the impact that they had on my life. I thought back to the days and nights, the joys and sorrows, the enchanting places and adventures I shared with many of these people whose faces still blaze in my memory like the glowing ashes of a flame. Thinking back and remembering all the things inspired me to write this story and reawakened my spirit of adventure, and there is still so much to look forward to.

AFTERWORD

Prime Minister Winston Churchill tasked
The Royal Navy instructed by
Admiral Earl Mountbatten,
Commander-in-Chief Mediterranean,
In helping to locate and retrieve
The comet wreckage, so that the cause
of the accident could be determined.

~

"The cost of the Comet mystery
Must be reckoned neither in money
Nor in Manpower."

Prime minister Winston Churchill, 1954

A LEGACY TO THE WORLD OF AVIATION

THE DE HAVILLAND COMET 1:
THE WORLD'S FIRST COMMERCIAL
(PASSENGER) JETLINER.

ABOUT THE AUTHOR

Elaine Baker was born in a small town in Rhodesia (now Zambia) and was educated in Natal, South Africa. Though her parents wished for her to become a secretary, she was entranced by the glamour of the burgeoning air travel business and relocated to London. There, she became a stewardess for BOAC and was soon flying all over the world.

As an ambassador for the airline, Ms. Baker helped usher in the Jet Age, making appearances and mixing with celebrities in the United States to help promote international flight. After leaving BOAC, she founded and operated her own dance academy. Then as a production accountant she worked with movie companies throughout the world.

Ms. Baker has lived in London for the past thirty years and continues to write and travel extensively.

Milton Keynes UK
Ingram Content Group UK Ltd.
UKHW051512070224
437434UK00011B/46/J